JOHN GLOAG ON INDUSTRIAL DESIGN

Volume 2

THE PLACE OF
GLASS IN BUILDING

THE PLACE OF
GLASS IN BUILDING

Edited by
JOHN GLOAG

Routledge
Taylor & Francis Group

LONDON AND NEW YORK

First published in 1942 by George Allen & Unwin Ltd.

This edition first published in 2023
by Routledge
4 Park Square, Milton Park, Abingdon, Oxon OX14 4RN

and by Routledge
605 Third Avenue, New York, NY 10158

Routledge is an imprint of the Taylor & Francis Group, an informa business

© 1942 John Gloag.

British Library Cataloguing in Publication Data
A catalogue record for this book is available from the British Library

ISBN: 978-1-032-36309-7 (Set)
ISBN: 978-1-032-36573-2 (Volume 2) (hbk)
ISBN: 978-1-032-36581-7 (Volume 2) (pbk)
ISBN: 978-1-003-33270-1 (Volume 2) (ebk)

DOI: 10.1201/9781003332701

Publisher's Note
The publisher has gone to great lengths to ensure the quality of this reprint but points out that some imperfections in the original copies may be apparent.

Disclaimer
The publisher has made every effort to trace copyright holders and would welcome correspondence from those they have been unable to trace.

THE PLACE OF
GLASS
IN BUILDING

Edited by JOHN GLOAG, Hon. A.R.I.B.A.

With contributions by

LIONEL B. BUDDEN, M.A., F.R.I.B.A. and G. A. JELLICOE, F.R.I.B.A.

GEORGE ALLEN & UNWIN LTD
40 MUSEUM STREET, LONDON, W.C.1

1943

First published in 1942

ACKNOWLEDGMENTS

For their help in the collecting of material for this book, and for their critical scrutiny of the technical information contained in these pages, I am indebted to the Research Departments of Pilkington Brothers Limited, of St. Helens, Lancashire, and Chance Brothers and Co., Limited, of Smethwick, Birmingham. I am grateful also to Professor Lionel B. Budden, of the School of Architecture of Liverpool University, for contributing his views on the study of materials in connection with architectural education, and to Mr. G. A. Jellicoe, F.R.I.B.A., for his observations on the technique of using Glass Bricks in contemporary domestic architecture. Finally, Mr. Philip Unwin has put me in his debt, for his enthusiasm and constructive help in the production of this book.

JOHN GLOAG

Printed in Great Britain for George Allen & Unwin Ltd. by The Shenval Press

CONTENTS

THE PLACE OF GLASS IN BUILDING

By John Gloag, HON. A.R.I.B.A.

THE primary function of glass in architecture is to allow enclosed space to be illuminated by daylight. But glass has become more than a convenient transparent material; it has entered into structural partnerships with other materials; it has acquired new forms, new properties and new functions during the last twenty-five years, and may now be described as a structural material. Glass qualified for that description when the progress of building technique made it possible for external walls to be relieved of the task of supporting floors and roofs, when buildings became vertebrates instead of crustaceans, and an internal skeleton of steel changed outside walls to mere skins.

It has been pointed out by Professor Walter Gropius, and proved by contemporary architects, that "New synthetic substances—steel, concrete, glass—are actively superseding the traditional raw materials of construction. Their rigidity and molecular density have made it possible to erect wide-spanned and all but transparent structures, for which the skill of previous ages was manifestly inadequate."[1]

When plate glass was first manufactured in large sheets it became possible, in conjunction with structural steelwork, to make a big, stone-faced building appear to float on a glass tank. There is no reason why all the display windows on the pavement level in shopping streets like Regent Street London, Princes Street Edinburgh, or Fifth Avenue New York, should not take the form of a continuous horizontal glass band, with heavy stonework beginning at the first floor, clipped on to the steel girders. If the streets shed their romantic stonework, the buildings could have continuous glass walls from the pavement to the roof-line, which in certain commercial and industrial buildings might be desirable. As the wall is now a skin, nothing save an archaic affection for bogus grandeur prevents it from being a light, decorative and appropriate

---

[1] *The New Architecture and the Bauhaus,* by Walter Gropius, translated by P. Morton Shand. (Faber & Faber, 1935.)

integument, adequately insulated to exclude cold and noise.

In his stimulating essay, *The Scientific Attitude*, Dr. C. H. Waddington has some significant things to say about the relations of architects with the inventors and fabricators of new materials. "Just as painters naturally associated with anatomists at the time when human bodies were first being systematically dissected and anatomy was made into a science, so architects nowadays, confronted with new building materials such as glass and reinforced concrete, find it necessary to keep in touch with physicists and engineers."[2] The properties of glass, considered in relation to other materials, are dealt with in an illuminating sentence, which condenses into sixty-five words the structural evolution of modern architecture. "Suddenly the modern architect is presented with materials, steel for taking tensions, reinforced concrete which takes both tensions and compressions, which enable him perfectly simply to roof far larger spaces than any of his ancestors could do with their most elaborate fan vaultings and domes, and completely to abolish the problem of the larger window, by making, if he wishes, the whole wall of glass."[3]

A "wall of glass" has a fascinating sound; transparent materials have always appealed to the imagination of designers; and great buildings that have relied partly upon daylight for the delineation of their form, are unforgettable. The Crystal Palace was not regarded as the progenitor of a structural revolution when first it glistened in Hyde Park: it was appreciated and admired as an almost miraculous achievement in transparency. W. R. Lethaby, in writing of the contribution of the French Gothic school to architecture, states that "a cathedral became a stone cage with films of stained glass suspended in the voids, a marvellous jewelled lantern."[4] He also described a structural revolution that took place in the fourteenth and fifteenth centuries, when "the wall gathered itself up into tense shafts and piers, from which

---

[2] *The Scientific Attitude,* Chapter IV, "Art Looks to Science," page 49. (Penguin Books, 1941.)
[3] *Ibid,* page 49.
[4] *Architecture,* by W. R. Lethaby, Chapter XII, page 209. (Home University Library, 1912.)

branched the ribs of the vault; the windows spread, to occupy the whole curtain of wall between the shafts, and in doing so almost inevitably became many-mullioned and traceried; the body thus became all post and space, a cage of stone." [5]

The result of this mediæval revolution was a wall of glass. The glazed stone cage was appropriate as a form for a cathedral; the modern glass wall or the great expanse of plate glass window are appropriate for industrial buildings or shops; but in domestic architecture, in civic buildings, in offices, the relative proportions of voids and solids must be determined by human considerations. The factory may be a well-lit, well-ventilated hive of industry, its function is mechanistic; but the home is not and should not be an expression of what the late Frank Pick has called "the wicked doctrine that the house is a machine . . ."

The transparent wall of glass has dangers and disadvantages when introduced in domestic architecture.

Large expanses of glass in a room increase the cost of heating in winter, and demand a heavy outlay on fabrics for covering the transparent wall at night. The need for privacy, the site of the house, or flat, and the character of the climate in these islands should determine the area of transparency in any room. No room should be denied abundant daylight, but "a wall of glass" may bring a surfeit, which compels the occupants to mask the window, and reduce its size to bearable proportions.

During the last twenty years much has been said about an "international style," which is presumed to arise from the logical and functional use of the materials that have impelled the structural revolution.

Julian Leathart has pointed out that: "There is, most probably, no such thing as a purely logical process in architectural design. In the most ruthless exemplification of the modern style, æsthetic intention is perceptible; the arrangement of window opening in relation to solid wall, the disposition and relationship of the component units of mass and line, and in

---

[5] *Ibid*, page 205.

the colour of walls, windows and doors, are to be found evi-
dences of some pre-conceived intention quite apart from the
formal utility of the structure."[6] As for the word "functional,"
it becomes meaningless unless national and climatic characteristics are accommodated by those who use the new materials.
Daylight in Southern England is different from Scottish,
Italian and American Pacific Coast daylight. It must be
remembered that "The amount of water vapour present in the
atmosphere, and the relative diffusing from sky and ground,
influences the quality of the light as well as its quantity."
Also, that "The efficiency of a window in England as a source
of light depends on the amount of sky framed within it."[7]

That beguiling phrase, "a wall of glass" must not be per-
mitted to inflict upon the British householder or flat-dweller
the chill publicity endured by fish in an aquarium. An
Englishman's house is his castle: how he lives in it is his own
private business, hence window curtains. It has been the
fashion to laugh at the way the Victorians cluttered up their
houses with all kinds of unnecessary frills, but the starched,
white, lace curtain was a highly functional instrument for
the diffusion of light. It has never been bettered for this
purpose, because it provides absolute control over the diffusion of the light available, and, if drawn on a dull day, can
give additional brightness to a room.

The function of glass is not only the transmission of daylight
to an interior, but the diffusion of light. The capacity for
diffusion does not correspond to the amount of transmission.
Transparent glass that admits light into a room is inferior in
diffusing capacity to a translucent glass. Transparent glass
enables you to see out: it does not make the most of the light
that it lets in: a window that used glass to the best advantage
would be glazed in the upper part with a diffusing pattern,
and with sheet or plate glass in the lower panes where vision
was desirable. Control of the amount of diffusion, when a

[6] *Style in Architecture*, by Julian Leathart, F.R.I.B.A. Pages 160-161. (Nelson & Sons, Ltd. 1940.)
[7] *Theory and Elements of Architecture*, by Robert Atkinson, F.R.I.B.A., and Hope Bagenal, A.R.I.B.A. Pages 25 and 349. (Ernest Benn Ltd. 1926.)

diffusing glass is used in the window, can be obtained by drawing opaque curtains to narrow down the area of diffusion.

A simple demonstration of the diffusing properties of various glasses (notably the glasses that are known under the generic term of "cathedral" glasses), may be staged with a magic lantern in a darkened room. The lantern throws a beam which illuminates a sharply defined section of wall. Anybody standing behind the lantern can see the beam and the illuminated area, but little else. (If a lot of men in evening dress were in the room, a dim, white haze would indicate the collective effect of their shirtfronts.) If a piece of one of the good diffusing glasses is held in front of the lens of the lantern, the beam will be diffused, so that the light is no longer concentrated, but is distributed throughout the room, making everything visible.

But glass as a structural material, is not just a window-filler. The technical progress in the manufacture of the material has been considerable during the last quarter of a century. Some astonishing things have happened to its character, which have changed the traditional conception of it as a fragile, brittle substance. For example, a glass is now available that can bend and twist under pressure and which possesses great heat-resisting powers. It is plate glass subjected after manufacture to a toughening process. When tried beyond its tested strength it disintegrates, breaking up into small, powdery crystals. Its character recalls the legend of the luckless inventor of unbreakable glass, who showed an example of his work to the Emperor Tiberius. As the invention seemed likely to upset vested interests, there was a prompt execution. The legend varies: sometimes the Emperor is Nero. S. Baring-Gould, in *The Tragedy of the Cæsars*, refutes the story, and describes it as "a folk-tale told everywhere, and at all times." He adds: "How this story originated we find from Pliny the Elder, who says that in the reign of Tiberius a sort of flexible glass was invented, but as it was thought that this discovery would injure the trade of the manufacturers of vessels in bronze and other metals, the workshop of the ductile glass-makers was closed and the manufacture stopped. This was clearly done by order of the senate. The blame of stopping the manufacture was transferred to the emperor and a cock-and-bull story added."[8]

Glass reinforced by wire-netting embedded in its core has been in use for many years. When it was made with a square mesh, with the intersections of the wire electrically welded, it acquired an agreeable, decorative character. But before mentioning any of the forms of glass that have contributed to the structural revolution in architecture, some facts should be given about the British glass industry, and a simple answer given to the question: How is glass made?

Glass is made by melting together sand, soda and limestone. Many glasses now in use for special purposes contain other constituents, sometimes in large proportions. One furnace can produce up to one hundred and sixty square feet of glass per minute, continuously day and night all the year round.

The major glass industry of Britain has a unique and most progressive character. In a book I wrote some years ago, on the development of design in relation to industrial materials, I attempted to summarise a few of the glass industry's characteristics, as follows:

"The manufacture of glass illustrates, perhaps more lucidly than any other section of industry, the effect of imagination upon the nature of a product. In the making of structural and decorative glass there is in this country a technical fecundity that has changed the whole character of glass as a material for building. This inventiveness is not sporadic; it is continuous, and it is cautious. In this branch of industry designers are often present at the councils of manufacturers. There are more working partnerships between technicians and designers than are suspected, and some of the great English glass-makers are demonstrating the practical value of such collaboration.

[8] *The Tragedy of the Cæsars*, Volume I, page 373. (Methuen & Co. 1893.) M. P. Charlesworth, in *Trade-Routes and Commerce of The Roman Empire*, states that "In the reign of Tiberius a factory for making glass was started close to Porta Capena, and though the legends about the invention of 'unbreakable glass' are absurd and untrue, there can be little doubt that they faithfully reflect the amazing progress which the manufacture of glass made in this period." Chapter III, page 52.

"Of most English manufacturers it may be said that they will never release for consumption any product that is still in its experimental stage. Their technical integrity, if it may be put that way, is unshakable. None of the revolutionary types of glass that has appeared during the last few years has been released casually when it was still in the bright young idea stage; none of them has been allowed near an architect's office until it has thoroughly satisfied its makers."[9]

New forms of glass produced in recent years perform the most unusual tasks. There is a glass that admits a high proportion of natural ultra-violet radiation, and these powers of admission are permanent. There are also glasses of orange and bluish-green hues which are used in jam factories, abattoirs, dairies and in some larders, which cure the fly nuisance in summer. Flies dislike the special tints which these glasses impart to daylight.

A multitude of decorative glasses exist, such as rectangular units of mirror mounted on fabric, which can be bent round columns and which can curve this way and that; and opaque, coloured glasses, and patterned glasses of various kinds. Opaque glass in ashlar sizes, or in large sheets, makes an admirable external skin for a building, for, like tiling, it can be kept clean by hosing. (*The Daily Express* buildings in Manchester and in Fleet Street, London, and the building illustrated on page 57, are examples of this external use.) There are glasses which can bend light, and turn it into dark corners. These are prismatic glasses which can direct light to the depths of the gloomiest passage, and give a generous share of natural illumination to basements and ground-floor rooms that would normally be dark. Glass bricks, which have been used in America for some years, are now made in England; from these, translucent walls can be built, just as brick walls are built, laid in courses. Mr. G. A. Jellicoe has contributed a special section to this book, on some aspects of the use of glass bricks in domestic building.

Glass in architecture to-day has ceased to be dependent

* *Industrial Art Explained*, by John Gloag, Chapter V, pages 137–138. (George Allen & Unwin Ltd., 1934.)

wholly upon other materials for its framework, It can now stand up independently as a structural material, and a new Crystal Palace, if one were built, could transcend Paxton's ideas as dramatically as the Transatlantic flying-boats transcend the visions of Columbus.

But it is not to facilitate the practical expression of such flights of fancy that this book is designed. It gives the facts about the physical properties, the possibilities and the limitations of the glasses in common use that are manufactured in Britain to-day, and deals also with the attributes of various specialised and decorative glasses. It is primarily a work of reference for architectural students, and it may be of interest to architects, for it is also a condensed survey of the progress that has been made in this structural and decorative material.

It is important that the products of a progressive industry should be used appropriately, and that their capacity should be appreciated. Progress continues. Research and experiment are not interrupted by external and irrelevant events. The most recent achievement of research is the welding of metal directly to glass. The revolutionary promise of this process will certainly not be ignored either by architects or the manufacturers of building materials.

Glass is a revolutionary material: it enables the architect to fabricate and manipulate something that is universal but impalpable, beyond price but without cost—daylight.

THE STUDY OF GLASS:

ITS PLACE IN ARCHITECTURAL EDUCATION

By Lionel B. Budden, F.R.I.B.A.

THE training of the architectural student to-day is a much more protracted and complex affair than it was when articled pupilage was accepted as providing an adequate professional education. So long as the problems of planning and construction with which the architect might be called upon to deal remained fairly constant in type and continued to develop by a process of gradual and orderly progression, so long as knowledge of the technical resources available could be easily mastered and so long as there was a general agreement about how things could and should be done, the initial education of the architect could be accomplished well enough through the agency of articled pupilage. But when those conditions no longer held, when the programmes of architecture ceased to be restricted to a familiar range of subjects, when new needs had necessarily to be embodied in new conceptions, when new structural materials and new applications of science had combined greatly to extend the field of technical knowledge and imaginative opportunity open to the architect, then schools of architecture alone could attempt to offer a basic training sufficiently comprehensive and systematic to meet the situation that had arisen.

Schools and academies of architecture are institutions of long standing; they existed in the Roman Empire and arose again in Europe during the Renaissance, to be further increased in number and extended in range to cover the whole field of architectural education in our own time. In the early stages of their development the teaching of many of the British schools tended to be academic in the sense that it was chiefly pre-occupied with architectural form, with composition, with the arrangement of shapes, stylistic elements and details. That phase has long since passed and a large portion of the curriculum of all schools of architecture providing courses leading to professional qualifications is now devoted to the study of contemporary building resources and tech-

15

nique. For it is generally appreciated that a knowledge of the physical properties of materials, old and new, their nature, behaviour, structural, formal and decorative possibilities must be part of the elementary equipment of the architect. He must not only know what has been done, he must understand why it has been done and what are the limitations that have controlled the way in which materials have been used. In short, he must learn what things are practicable and what appropriate.

Amongst the materials whose use has been greatly extended in the present century is glass; so much so that it is often included with concrete amongst the new materials now at the disposal of the architect. In point of fact neither glass nor concrete are new materials. Concrete was employed by the Romans to provide the structural core of most of their great buildings; the secret of its composition was lost when the Roman Empire collapsed and has merely been re-discovered. Similarly glass was used throughout the later centuries of the Middle Ages upon an increasing scale and with increasing skill. The range of its use has in recent years been enormously expanded as a result of scientific research and experiment, but that does not make it a new material.

This point is stressed here because the idea of newness which has been popularised in connection with concrete and glass can and does undoubtedly tend to produce illusions in the mind of the embryonic architectural student; and these illusions have at an early stage to be dispelled. If a material with great structural or decorative potentialities is thought of as new it is apt to be invested with magic virtues of an extraordinary kind. Because glass is welcomed with enthusiasm as a new medium of architectural expression, extravagant assumptions are quite commonly made about its capacity to do the work of all other materials and to embody satisfactorily any and every sort of programme. Much patient persuasion is sometimes needed to convince an ardent but immature convert to "The Glass Age" that few English families would really enjoy living in a house that was virtually a conservatory; that a music room whose walls, ceiling

16

and floor were of glass would present shattering acoustic problems which could only be solved by a change of material; and that heating costs in this country are not as a rule a negligible item in domestic economy.

To enable the student to use the resources of glass rightly in architectural design, he must know something about its physical constitution and properties; he must become familiar with the various types and forms of glass which are available, both structural and decorative; and he must learn about those methods of fixing the material which have proved to be satisfactory in practice; in short he must acquire from lectures, from studio instruction, and from his own investigations the kind of information which is assembled in this book and which in schools of architecture possessing a gallery of materials may be most valuably supplemented by the actual observation and handling of samples. But beyond all this he must study intelligently the way in which glass has been used with other materials to meet the needs—some constant, some changing—of architectural programmes themselves. Here actual buildings and illustrations of buildings will be the proper subjects of his attention. If his examples are well chosen he will discover that the exercise of practical sense is as necessary in the appropriate treatment of glass as in the treatment of any other architectural material; and that to observe this elementary condition need not mean that imaginative qualities must be sacrificed. The achievements which glass, used in conjunction with reinforced concrete or steel, has made possible in the field of industrial and commercial architecture, in the design of hospitals, schools and other buildings, are now familiar to us all. So also is the wide range of the decorative effects in colour, pattern and texture which glass has in recent years been able to afford. Research and experiment are being continued and still more possibilities, structural as well as decorative, may soon be revealed. Of all this architectural education must continue to take account; for whilst the principles of architecture remain permanent, its technical resources change and the knowledge relating to those resources must influence both teaching and practice.

USE OF GLASS IN SMALL STANDARD HOUSES

By G. A. Jellicoe, F.R.I.B.A.

UNTIL recently the consideration of glass in contemporary standard houses was confined to its relation to window areas only. Minimum standards of window space were established by bye-laws, and these ensured the minimum requirements of light. The tendency of window design was if anything to increase in size and let in more and more light. To suit these requirements and those of mass production, window manufacturers produced a standard window design consisting mainly of a reasonable sized unit that could be added to according to the wish of each designer. The use of glass was almost wholly confined to this one pane, despite the occasional experiments of architects here and there.

Except perhaps for the use of water in garden design, that of glass in house design lends itself to the play of imagination more than any other material. Its function in the past has been to let in light and let out vision. It excludes wind and rain, and less efficiently changes in temperature. To a certain extent also it may preclude excessive direct light; Mr. Gloag points out in his introduction that the Nottingham lace curtains diffuse light and actually increase its value, and glass may be called upon to do the same. Another function may well be that it will let out light still without letting in vision.

These are broadly the purposes of glass, and give the clue to its decorative qualities. Possibly the first of these is its essential preciousness. It is a dainty material set in the stronger and coarser materials of walling. It concerns light, and the play of reflection upon its surface can be as lively as the sparkle of light on falling water. Before good glass was made, the Crown glass flashed and sparkled in the Georgian façade. With the unevenness of the old glass overcome, much of the pleasure of glass has departed. But it may be returned in a different form and under different conditions.

The diagrams show the use of glass in a standard house. They illustrate how glass may be used economically and decoratively. The glass bricks on each side of the porch are set in a surround that gives a sense of protection to so precious a material and thereby tends to emphasize the fastidiousness of the glass itself. On plan the divided light gives an even distribution in the hall, one side lighting the downstairs passage, and the second lighting the staircase. The hall is of sufficient proportions for the quality of the light to be appreciated. From the outside, although it is not possible to experiment with this in war-time, it is anticipated that the light in the hall will throw a glow through the glass bricks that will make a porch light unnecessary and create a sense of welcome after dark. On a sunny day from across the street it is attractive to watch the sun glinting on the edges of the glass, but an unexpected discovery has been the effect of the sun upon a façade of houses facing north. When one of the internal doors is left ajar or open, the glass bricks pick up and diffuse over their own surface the patches of sun beyond. When, therefore, the south side of the street is in shade and seen against the sun, the porches are often glowing with light.

The sense of protection to door and glass

Internal distribution of light on floor

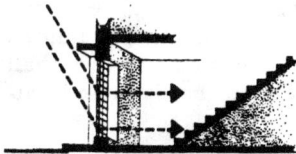

The windows of this house are of the usual standard type,

Internal distribution of light in section

but they have been set in a reconstructed stone frame that preserves them from the coarseness of the brick. The windows themselves are set either deeply in the sill, or are brought out to the surface, according to their position. The variety given to this in perspective is interesting. The glass set forward

Recessed windows in concrete frame

picks up the reflections of the sky, for there are no shadows thrown upon it. The glass set back loses its sparkle, because of

Flush windows in concrete frame

deep shadow, but the void thus made emphasizes the solidity of the building. It may well be that the study of street reflections will become as engrossing as those made of water by the 18th century landscape designer Humphrey Repton.

Another point about the use of glass in this design is the contrast between the scale of the glass bricks and that of the window panes. There is a purely two-dimensional play on design that helps to give intricacy; the total façade is not wholly comprehended at one glance.

The example analysed is very simple, but it seems possible that provided the qualities of glass are fully understood, this material may have a considerable future in standard house design. The pitfall to avoid is exaggeration which so often comes from the discovery of a new material. The period between the two wars seems to be the one where in all our history the least consideration has been given to the sense of protection required by external materials. If in our building we provide this sense of protection not only to the human being but to the structure of the building itself, then we shall retain that human quality in house design that is the inheritance of this country from its past.

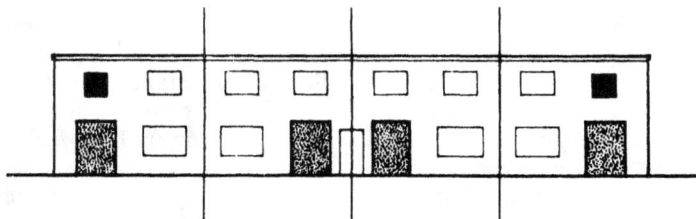

DIAGRAM OF ELEVATIONAL GLASS PATTERN
Black: recessed windows. Grey: porches. White: flush windows

This quality also concerns the craftsman. More and more the factory for prefabrication and the large machine on the site will both tend to eliminate the human element in building. This is as it should be, for the process gives a greater all-round efficiency. But the man on the job will never be eliminated in this country, and his individuality can still be combined with the machine element to create a personal building. Thus the study of modern brickwork in relation to the mass-produced house has scarcely begun, but there is already sufficient proof to show that one gang of bricklayers can bring delight into a wall surface, where another in an adjoining district working upon exactly the same specification will make a surface that is æsthetically dead. So it can be with glass.

PREPARING SPECIFICATIONS

IN preparing specifications for glazing, it is advisable to include a general clause to this effect:

All glass to be of the type, quality, and substance specified, and to be of British manufacture (unless of course a specific glass of foreign manufacture is specified by name). The glazier must be prepared to produce at the completion of the job invoice or voucher from the manufacturer to show that the glass supplied is to the standard specified.

Glasses should always be described by the recognised trade terms, thicknesses and qualities (see B.S.I. Specification 952/1941).

Under the sections devoted to Sheet and Plate Glass, there are diagrams which enable the maximum safe glazing sizes to be determined, under normal conditions. Where glazing is proposed for sites where there are abnormal conditions, it is advisable to obtain technical advice from the manufacturers of the glass to be used, regarding wind-pressure and so forth. The curves for Sheet Glass have been calculated for the minimum thicknesses supplied under the tolerances allowed in the respective nominal substances. The curves for Plate Glass have been calculated for the exact thicknesses by which the respective substances are designated.

The sizes are calculated on the basis of the strength of the glass and are not related in any way to the maximum manufacturing sizes, with which they must not be confused.

SHEET GLASS

THE MANUFACTURE OF MODERN (FLAT DRAWN) SHEET GLASS
Glass, formed by the reaction at high temperature, of sand, soda, lime and certain other ingredients, is made in a large gas-fired tank furnace, by heating to a temperature around 1,500° C. The heating is confined usually to one end of the tank, and the glass formed at this end—the melting end—flows down to the opposite or working end of the tank, gradually cooling as it moves away from the source of heat.

To form the molten glass into a sheet, it first passes from the tank into a drawing kiln, a relatively small extension to the tank and separated from it above the level of the glass surface by a "tweel" and "shut-off." The shut-off is a block of refractory material which floats on the surface of the glass. The tweel is a slab of refractory material suspended from above and lowered until it rests on the shut-off and completes the seal between the tank atmosphere and that of the kiln.

After entering the drawing kiln underneath the shut-off, the molten glass flows round either side of a submerged clay block of special design, known as the "drawbar," and is then drawn up in a continuous ribbon. The power needed to draw the ribbon is supplied by a series of electrically driven, asbestos-covered rollers mounted in pairs in a cast-iron tower, situated above and parallel to the length of the drawbar. To start this process an iron grille known as the "bait" is lowered between the tower rollers into the molten glass. When it has remained there for a short period the molten glass sticks to the bait which is then slowly lifted, drawing behind it a ribbon of glass. When the leading edge has passed through the first few rollers the bait is cracked off, allowing the glass ribbon to be drawn up into the tower.

The success of the process lies in the provision of devices for maintaining the width of the ribbon of glass while being drawn, since, being in a plastic condition, there is a marked tendency for the glass to "waist," that is, shrinking to a mere thread. The usual mechanism employed for this purpose consists of a fork and a pair of knurled air-cooled steel rollers. The fork—a slightly curved steel plate with a machined slot in it, is placed just above the level of the molten glass in the kiln, so that the edge of the ribbon draws through the slot, and is then engaged by the knurled rollers. The two rollers are pressed towards each other so as to grip the edge of the glass firmly. They cool the edge of the glass sheet, thereby preventing subsequent "waisting," and are usually driven independently of the tower rollers.

Facing the glass, at a position just above the level of the molten glass in the kiln, are water-cooled steel boxes, which help to solidify the ribbon when formed. Once it has been formed it can be drawn continuously for four or five weeks, until it becomes necessary to re-heat the kiln.

As the glass is drawn up the annealing tower, it gradually cools off, and at a height of about 40 feet above the kiln, it is sufficiently cold to be cut off. This is done by putting a wheel-cut on the rear surface of the glass, a mechanically driven cutter operated automatically when the glass reaches a certain pre-determined height, producing the cut. The sheet of glass is then pulled by hand away from the cut, which opens and enables the cut sheet to be removed and trimmed. This trimming consists of the removal of the edges, these bearing the marks of the knurled rollers. The glass lost by this edge trimming is returned to the tank and re-melted. From this stage the sheet glass is sent into the warehouses.

Sheet Glass is the most commonly used glass for general glazing. It is fire-finished, and in consequence the two surfaces are never perfectly flat or parallel. This accounts for a certain amount of distortion of vision and reflection which is unavoidable with this type of glass.

Sheet Glass—Flat Drawn Process: The sheet being cut off at the top of the annealing tower into the required lengths by an electrically operated cutter.

Sheet Glass is made in various thicknesses which are described in terms of ounces per square foot.

Weights and Thicknesses

18 oz.	...	approx.	1/12″	26 oz.	...	approx. 1/8″
24 oz.	...	,,	1/10″	32 oz.	...	,, 5/32″

Qualities and Character

Each of the thicknesses is supplied in three recognised standard qualities, i.e.:

ORDINARY GLAZING QUALITY

(Referred to as O.Q.) Suitable for general glazing purposes.

SELECTED GLAZING QUALITY

(Referred to as S.Q.) For glazing work requiring a selected glass above the ordinary glazing quality.

SPECIAL SELECTED QUALITY

(Referred to as S.S.Q.) For high-grade work where a super-fine glass is required.

Light Transmission

The absorption of light is so small that although reflections account for a loss of over 8 per cent, the overall transmission is approximately 90 per cent of the available light.

Safe Glazing Sizes

Curves corresponding with each thickness have been prepared to show the maximum safe glazing sizes in that substance in conditions of exposure not exceeding a wind pressure of 15 lb. per sq. ft. (68.5 m.p.h. wind velocity). Any square, rectangular, or circular size, that

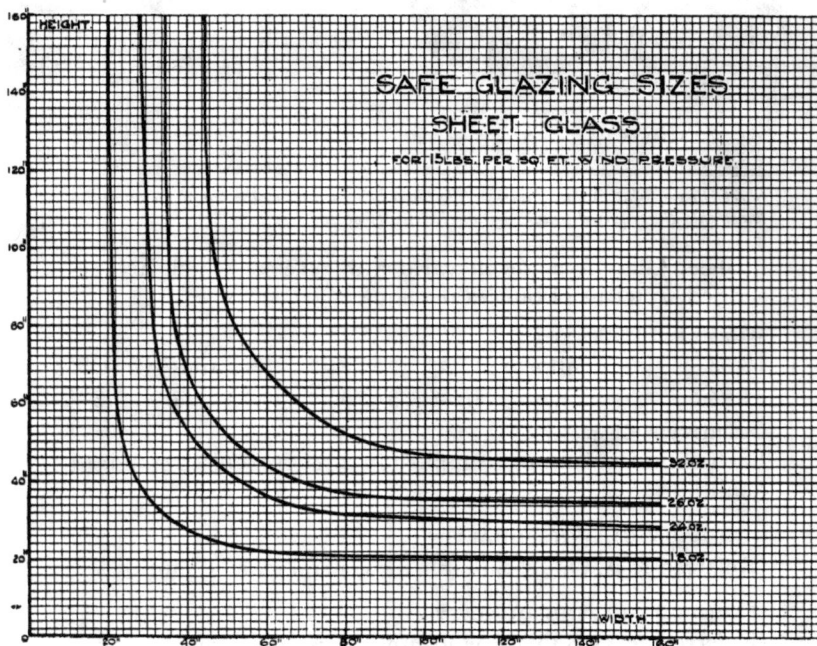

SAFE GLAZING SIZES
SHEET GLASS
FOR 15 LBS. PER SQ. FT. WIND PRESSURE

can be fitted under the curve corresponding to each substance, fulfils the L.C.C. requirements. For abnormal sites—e.g. when a window forms a wind-pocket or when the building is in an unusually exposed position, the manufacturers should be consulted.

Examples

32 oz.—100″ × 47″, 80″ × 52″, 64″ square, 41″ high × any width, 41″ wide × any height, etc.

26 oz.—90″ × 36″, 68″ × 40″, 51″ square, 32″ high × any width, 32″ wide × any height, etc.

24 oz.—80″ × 32″, 60″ × 36″ 46″ square, 28″ high × any width, 28″ wide × any height, etc.

18 oz.—60″ × 22″, 48″ × 24″, 33″ square, 20″ high × any width, 20″ wide × any height, etc.

USES

Sheet Glass is commonly used in factories, housing estates, and for horticultural purposes. The weight and size of Sheet Glass used in any job should be governed by a consideration of the information given above. For example, for small windows in a double-hung sash used on a housing estate, the safe glazing size suggests 18 oz., but 24 oz. is recommended to provide the additional safety factor. Generally speaking, it is advisable to use a slightly thicker glass than that required to meet the normal safety factor.

POLISHED PLATE GLASS

THE MANUFACTURE OF POLISHED PLATE GLASS

Polished Plate Glass, after "casting" or "forming," has its surfaces mechanically ground and polished. It is melted and refined in the same type of tank furnace as Sheet Glass, and the chemical composition is similar, except that for Polished Plate the materials are selected and refined. The raw materials are stored in silos in the mixing room, and weighed quantities of each are thoroughly mixed and the resulting material is known as "frit." "Batches" of frit are fed into the melting end of the tank furnace at intervals to balance the quantity of glass withdrawn at the working end, to ensure that the amount of molten glass in the tank remains constant. When the molten glass reaches the delivery end of the tank furnace, it is homogeneous and free from bubbles, and, at a temperature of over 1,000°C. passes to a pair of water-cooled steel casting rollers, between which it is rolled into a continuous ribbon. The width of the glass ribbon depends upon the amount fed to the rollers; the thickness can be varied between 4 mm. and 30 mm. by adjusting the gap between the rollers. The glass ribbon is passed through a long, horizontal annealing lehr, being supported upon and carried forward by steel rollers. Temperature distribution in this lehr is such that differences in temperature in the glass ribbon are equalised, and the glass cools to atmospheric temperature without having induced in it stress sufficient to cause warp or to affect the cutting.

When it emerges from the lehr, the glass has rough, unattractive surfaces which must be removed by grinding and polishing to obtain the highly polished surfaces associated with Plate Glass. So from the lehr the ribbon of glass travels forward to the machine which grinds and polishes the surfaces simultaneously.

This process of grinding and polishing both surfaces of a moving ribbon is the latest development in Plate Glass manufacture. Each grinding head above and below the ribbon has a vertical motor-driven spindle carrying a cast iron disc in which there are grooves to ensure even distribution of the sand and water. The first pair of discs is fed with coarse sand, while the succeeding discs receive sand which becomes progressively finer. During the grinding process about $\frac{1}{2}$ mm. of glass is removed from each surface.

The polishing machine has a large number of heads similar to the grinder, but instead of iron discs each head carries a number of felt discs which are arranged to rub the glass in all directions. A mixture of rouge and water is fed on to these discs and converts the dull "frosted" finish left by the grinders into highly polished surfaces.

Twin Plate is the name given to British Plate Glass made by this process and it has many advantages over that made by other methods. From the user's point of view the main advantage is the flatness and parallelism of the two surfaces. The combined effect of these is to give a glass free from distortion and wave, eminently suitable for glazing, the manufacture of mirrors, and in all places where glass of good quality is desired. Twin Plate is being used in instruments and other special applications where formerly optically worked glass was employed.

The grinding and polishing machine just described is the only one of its kind in the world and is the result of many years of research and experimental work. It is developed from the previous type in which the glass is laid on a continuously moving table which passes beneath grinding and polishing heads operating on the upper surfaces of the glass only, and the glass must therefore be re-laid for grinding and polishing the other side. When this type of grinding and polishing machine is used, the ribbon of rough cast glass emerging from the annealing lehr is cut into suitable lengths for handling and laying. Plate Glass made by this process, though of excellent quality, has not the freedom from wave and distortion which characterises Twin Plate.

When the glass leaves the polishing machines, and before it reaches the warehouse, it is cleaned by passing it through a washing machine. In the warehouse, the glass is examined for glass and surface defects, under critical lighting conditions, before it is cut. For cutting, diamonds, ground to a suitable shape, are used; after the cut has been made, a thin wooden lath is placed under the cut, between the glass and the cutting table, and the glass snapped. If this is done skilfully, the break will follow the line of the diamond cut. For transport, the glass is packed vertically in stout wooden cases lined with straw, each plate being protected from its neighbour by interleaving with paper.

Polished Plate Glass has two surfaces ground, smoothed and polished, the object being to render the surfaces flat and parallel, and thus to provide clear and undistorted vision and reflection.

Thicknesses

Ordinary range, $\frac{1}{8}$" to $1\frac{1}{4}$" and up to $1\frac{1}{2}$" if desired. The normal substances supplied, unless otherwise stated, are approximately $\frac{1}{4}$". If a substance other than $\frac{1}{4}$" is required, it must be stated. It should be noted that glass thinner than $\frac{1}{4}$" is more costly, because, to arrive at this reduced thickness, other processes are involved.

Qualities and Character

Plate Glass is characterised by its flat surface and high polish. In its normal thickness of $\frac{1}{4}$" it is rather more than twice as strong as the thickest Sheet Glass ordinarily used for glazing windows (24 oz. 1/10"), and, being thicker, it offers higher thermal and sound insulation. Polished Plate Glass is supplied in three qualities :

G.G. = General Glazing Quality
S.G. = Selected Glazing Quality
S.Q. = Silvering Quality

Original type of Continuous Grinding and Polishing Machine used in the manufacture of Polished Plate Glass.

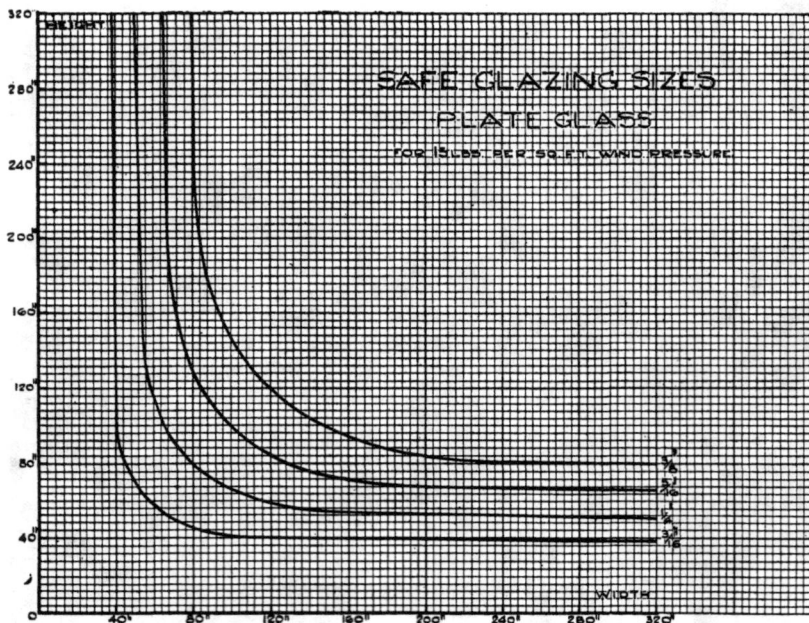

SAFE GLAZING SIZES
PLATE GLASS
FOR 15 LBS. PER SQ. FT. WIND PRESSURE

A.—*Curves corresponding with each thickness have been prepared to show the maximum safe glazing sizes in that substance in conditions of exposure not exceeding a wind pressure of 15 lb. per sq. ft. (68.5 m.p.h. wind velocity) Any square, rectangular, or circular size that can be fitted under the curve corresponding to each substance fulfils the L.C.C. requirements. For abnormal sites—e.g., when a window forms a wind-pocket or when the building is in an unusually exposed position, the manufacturers should be consulted.*

Light Transmission

The light absorption is so small that, even though 8 per cent is allowed for reflection, the total transmission is approximately 90 per cent of the available light.

USES

Polished Plate Glass should be used for public, commercial and domestic buildings, hotels, hospitals, schools and office blocks, etc.

32

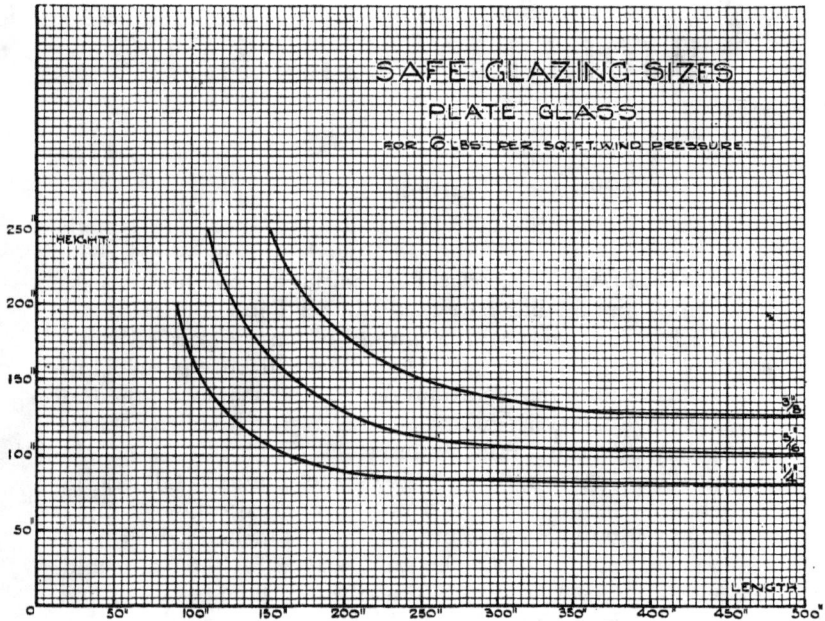

B.—*Any square, rectangular, or circular size that can be fitted under the curve corresponding to each substance, may be safely glazed in that substance at ground level in a normal town environment, with wind-pressure not exceeding 6 lb. per sq. ft. (42 m.p.h. wind velocity).*

For shop windows, display cases and in all instances where undistorted vision is required, Plate Glass is essential. Its high quality greatly enhances the exterior appearance of buildings.

ROLLED GLASSES

THE CONTINUOUS ROLLED PROCESS

There are three distinct types of Rolled Glass: (1) Rough Cast, (2) Cathedral and Figured Rolled, and (3) Wired.

(1) *Rough Cast:* The surfaces are not highly polished, nor do they carry any regular pattern. Plain (or Ribbed) Rolled has one surface impressed with a pattern of narrow parallel ribs.

(2) *Cathedral and Figured Rolled:* These have one side impressed with patterns of various depths and shapes. Cathedral Glasses have faint, irregular patterns. Figured Rolled Glasses have deeper, more formal patterns.

(3) *Wired:* Rolled, Rough Cast and Figured Glasses in which wire netting is inserted during the process of rolling.

In the continuous rolled process, glass is melted and refined in tank furnaces in much the same way as for the Sheet and Plate Glass processes already described. It then passes to one of several different types of machine, according to the sort of Rolled Glass it is desired to make.

The feature common to all machines making Rolled and Wired Glasses is the formation of the sheet by extrusion between two rollers, mounted parallel to each other, at a distance determined by the thickness of the sheet to be made. The molten glass is directed on to these rollers by allowing it to overflow from the tank through a specially designed refractory spout, the amount being controlled by an adjustable "tweel". This tweel is a clay block which can be lowered into the stream of glass and alters the effective depth of glass flowing to the spout. The glass flowing on to the backs of these rollers is forced, by their rotation, through the space between them, whence it issues as a continuous ribbon. This sheet, or ribbon of glass, is being formed continuously, and is taken by a driven roller conveyor through a long, horizontal tunnel or

A Continuous Glass Rolling Machine.

lehr. This lehr is heated, and the temperatures along its length are so adjusted that the glass passing through gradually solidifies and cools down to normal atmospheric temperature. When cool enough, it emerges from the lehr and is cut off in lengths suitable for transfer to the warehouses.

In the manufacture of Rough Cast, both rollers may be without pattern. Occasionally one is imprinted with a shallow, disruptive pattern. For Cathedral and Figured Rolled Glasses, one roller has a pattern cut in its surface, which imprints itself on the glass as it is formed. Usually the bottom roller is patterned, so that the glass emerging from the lehr has a smooth face upwards to facilitate cutting.

Sizes, Thicknesses, etc.

Nominal Thickness	Approx. weight per sq. ft. lbs. ozs.		Manufacturing sizes	Light transmission (approx.)
Rough Cast $\frac{3}{16}''$	2	10	$120'' \times 46''$ or $130'' \times 26''$	Diffused light for
Double Rolled $\frac{1}{4}''$	3	6	$120'' \times 48''$ or $144'' \times 26''$	$\frac{1}{4}''$ thickness: 80%
$\frac{3}{8}''$	5	1	$110'' \times 48''$ or $120'' \times 26''$	of the available light.
Plain Rolled $\frac{1}{8}''$	1	3	$120'' \times 42''$	Diffused light for
$\frac{3}{16}''$	2	10	$120'' \times 46''$ or $130'' \times 26''$	$\frac{1}{4}''$ thickness: 80%
$\frac{1}{4}''$	3	6	$120'' \times 46''$ or $144'' \times 26''$	of the available light.

Glazing Sizes

This type of glass is largely used for roof and factory lighting
in patent glazing bars, either vertical or horizontal, and in
suitable sizes specified by the trade.

USES

Rough Cast Double Rolled is used for skylights and roofing in
public buildings, warehouses, factories, workshops, etc., when
the extra protection afforded by Wired Glass is not considered
necessary. It is also used for vertical glazing in factories, warehouses, etc., where a transparent glass is not required.

Plain Rolled: Uses similar to Rough Cast Double Rolled.
The narrow parallel ribs have the effect of diffusing the light
and also reducing direct glare from the sun.

Figured Rolled Glass and Cathedral Glass come strictly into
the same category. Each is a Rolled Glass, one surface of
which has a definite pattern (Figured) or texture (Cathedral)
obscuring vision completely or partially, according to the
depth and configuration of the texture or pattern. The glass
may be tinted or untinted.

Figured Rolled and Cathedral Glasses are largely used for
partitions in offices and warehouses, when direct vision
is not desired, and also for window glazing of warehouses,
factories, etc., where Wired Glass or the heavier types of
Rolled Glass are not considered necessary.

The uses to which Figured Rolled and Cathedral Glasses are
put, will be governed by the considerations shown on page 45.

Rough Cast Double Rolled.   *Both surfaces of irregular texture, due to contact with rollers.*

Plain Rolled.   *One surface impressed with a pattern consisting of narrow parallel ribs (19 to the inch) while the other surface is flat*

Clear Cathedral.

Clouded Cathedral.

Double Rolled Cathedral.

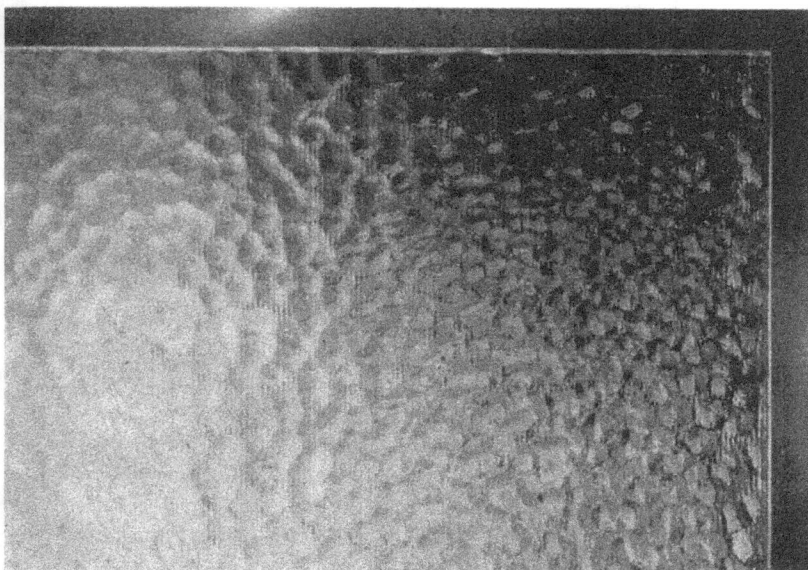

Rimpled Cathedral.

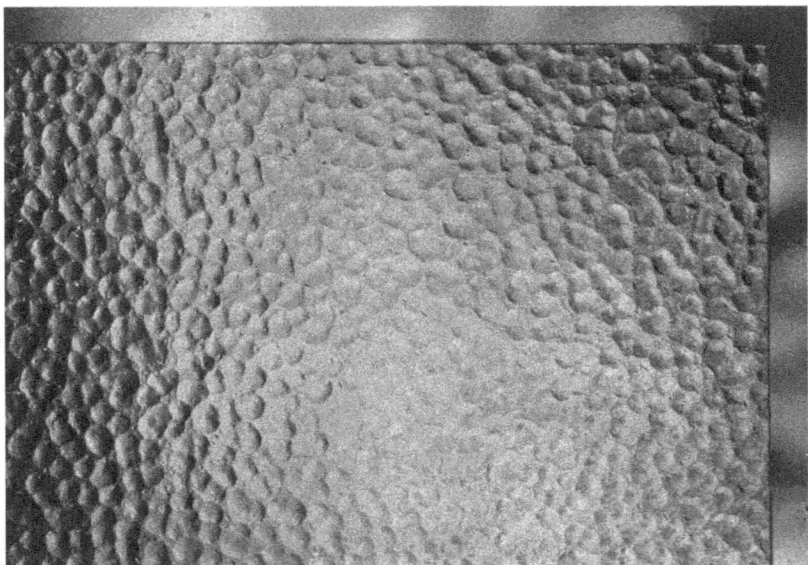

Mottled Cathedral.

Stippled Cathedral.

Hammered Cathedral No. 2.

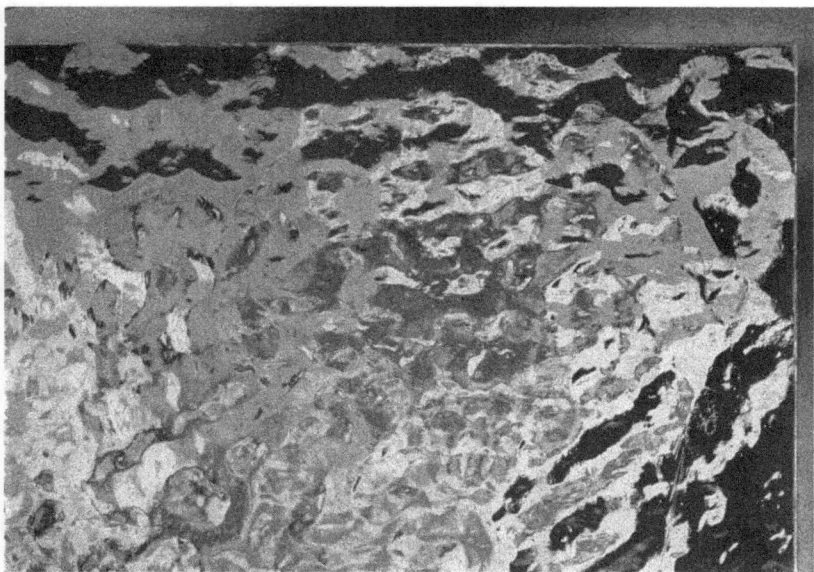

Glasgow Hammered Cathedral.

Arctic.

Majestic.

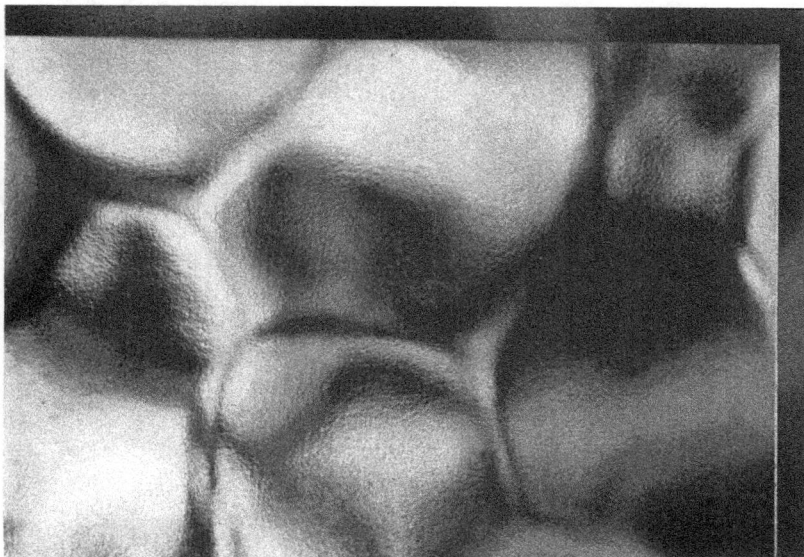

Large Flemish.

Pinhead Morocco.

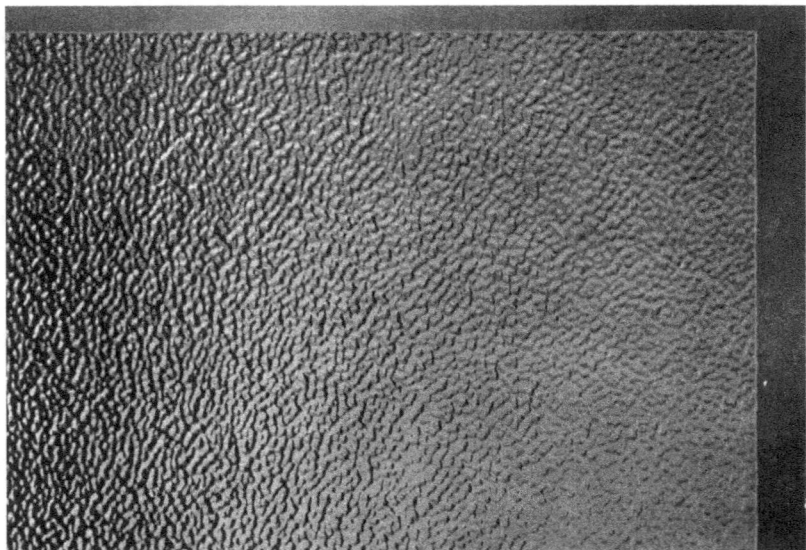

Stippolyte.

Kaleidoscope.

Sizes, Thicknesses, etc.

NON-FORMAL TEXTURE *The textured surface just gives sufficient obscurity to prevent clear vision through the glass*	Light Transmission	Normal Manufacturing Sizes	Thicknesses	Approx. Weight per sq. ft.	Tints Available
CLEAR CATHEDRAL	About 85% of the available light, with little diffusion	White 120"×48" Tinted 100"×36" or 90"×42"	$\frac{1}{8}$" $\frac{3}{16}$" $\frac{1}{4}$"	1$\frac{1}{2}$ lbs. 2$\frac{1}{2}$ lbs. 3$\frac{1}{2}$ lbs.	Available in 13 standard tints, and a comprehensive range of intermediate shades
CLOUDED CATHEDRAL (similar surface to Plain Rolled, but clouded instead of bright)	About 85% of the available light, with little diffusion	White 120"×48" Tinted 100"×36" or 90"×42"	$\frac{1}{8}$" only $\frac{3}{16}$" and $\frac{1}{4}$" can be manufactured if required	1$\frac{1}{2}$ lbs.	White only available at present, but could be supplied if required in 13 standard tints, and a comprehensive range of intermediate shades
DOUBLE ROLLED CATHEDRAL	About 85% of the available light, with little diffusion	White 120"×48" Tinted 100"×36" or 90"×42"	$\frac{1}{8}$" $\frac{3}{16}$" $\frac{1}{4}$"	1$\frac{1}{2}$ lbs. 2$\frac{1}{2}$ lbs. 3$\frac{1}{2}$ lbs.	Available in 13 standard tints, and a comprehensive range of intermediate shades
RIMPLED CATHEDRAL (similar to Double Rolled but with a slightly more pronounced pattern)	About 85% of the available light, with little diffusion	White 120"×48" Tinted 100"×36" or 90"×42"	$\frac{1}{8}$" $\frac{3}{16}$" $\frac{1}{4}$"	1$\frac{1}{2}$ lbs. 2$\frac{1}{2}$ lbs. 3$\frac{1}{2}$ lbs.	Available in 13 standard tints, and a comprehensive range of intermediate shades
MOTTLED CATHEDRAL (Having a slightly embossed pattern to give extra diffusion)	About 85% of the available light	White 120"×42" Tinted 100"×36" or 90"×40"	$\frac{1}{8}$" $\frac{3}{16}$" $\frac{1}{4}$"	1$\frac{1}{2}$ lbs. 2$\frac{1}{2}$ lbs. 3$\frac{1}{2}$ lbs.	Available in 13 standard tints, and a comprehensive range of intermediate shades
STIPPLED CATHEDRAL (Similar to Double Rolled with a slightly more diffusing surface)	About 85% of the available light	White 120"×42" Tinted 100"×36" or 90"×40"	$\frac{1}{8}$" $\frac{3}{16}$" $\frac{1}{4}$"	1$\frac{1}{2}$ lbs. 2$\frac{1}{2}$ lbs. 3$\frac{1}{2}$ lbs.	Available in 13 standard tints, and a comprehensive range of intermediate shades

45

	Light Transmission	Normal Manufacturing Sizes	Thicknesses	Approx. Weight per sq. ft.	Tints Available
SEMI-FORMAL TEXTURE *On one surface a slight semi-formal pattern is impressed giving brightness to the appearance of the glass. Direct vision is partly obscured*					
HAMMERED CATHEDRAL No. 2 (small)	80% to 85% of the available light, with very little diffusion	White 120"×48" Tinted 100"×36" or 90"×42"	$\frac{1}{8}''$ $\frac{3}{16}''$ $\frac{1}{4}''$	1½ lbs. 2½ lbs. 3½ lbs.	Available in 13 standard tints, and a comprehensive range of intermediate shades
GLASGOW HAMMERED CATHEDRAL (Similar to No. 2 Hammered but with a slightly less defined pattern)	About 85% of the available light	White 120"×42" Tinted 100"×30" or 90"×42"	$\frac{1}{8}''$ $\frac{3}{16}''$ $\frac{1}{4}''$	1⅛ lbs. 2½ lbs. 3½ lbs.	Available in 13 standard tints, and a comprehensive range of intermediate shades
FORMAL PATTERNS *A deeply impressed pattern giving a high degree of brightness to the glass. Direct vision through the glass is almost obscured*					
ARCTIC (large)	80% to 85% of the available light, with considerable diffusion	White 120"×48" Tinted 100"×36" or 90"×42"	$\frac{1}{8}''$ $\frac{1}{4}''$	1½ lbs. 3½ lbs.	Available in 13 standard tints
MAJESTIC	75% to 85% of the available light, with considerable diffusion	White 100"×42" Tinted 100"×36" or 90"×42"	$\frac{1}{8}''$ $\frac{1}{4}''$	1½ lbs. 3½ lbs.	Available in 13 standard tints
LARGE FLEMISH	82%–87% of the available light	White 120"×42" Tinted 100"×36" or 90"×40"	$\frac{1}{8}''$ $\frac{3}{16}''$ $\frac{1}{4}''$	1½ lbs. 2½ lbs. 3½ lbs.	Available in a limited number of tints including pastel shades

	Light Transmission	Normal Manufacturing Sizes	Thicknesses	Approx. Weight per sq. ft.	Tints Available
DIFFUSED PATTERNS *A deeply impressed pattern giving a high degree of brightness to the glass. Direct vision is completely obscured, with very little loss of light*					
PINHEAD MOROCCO	80% to 85% of the available light with almost complete diffusion	White 100″×47″ Tinted 100″×36″ or 90″×42″	$\frac{1}{8}$″ $\frac{1}{4}$″	1$\frac{1}{2}$ lbs. 3$\frac{1}{2}$ lbs.	Available in 13 standard tints
STIPPOLYTE	About 85% of the available light, with almost complete diffusion	White 120″×42″ Tinted 100″×36″ or 90″×40″	$\frac{1}{8}$″ $\frac{1}{4}$″	1$\frac{1}{2}$ lbs. 3$\frac{1}{2}$ lbs.	Available in 13 standard tints
COMPLETE DIFFUSION AND OBSCURATION PATTERNS *A deep geometric pattern is impressed, giving a high degree of brightness to the appearance of the glass, and complete obscuration with very little loss of light*					
KALEIDOSCOPE	About 75% of the available light, with perfect diffusion	White 120″×42″ Tinted 100″×36″ or 90″×42″	$\frac{1}{8}$″ $\frac{1}{4}$″	1$\frac{1}{2}$ lbs. 3$\frac{1}{2}$ lbs.	Available in 13 standard tints

The Glasses enumerated in the foregoing pages do not represent the full range of patterns available, but have been selected as being representative of their particular groups.

47

Wired Glasses: The glass is reinforced by a wire mesh embedded in the middle of the glass. Valuable as a safeguard against accident, burglary, and as an efficient fire retardative.

Sizes, Thicknesses, etc.

Translucent Types	Light Transmission	Manufacturing Sizes	Approx. Thickness	Approx. weight per sq. ft.
*Wired Cast	About 80% of the available light	120" × 40" or 144" × 26"	¼"	3¼ lb.
*Georgian Wired Rough Cast	About 80% of the available light	120" × 40" or 144" × 26"	¼"	3¼ lb.
Wired Arctic	About 80% of the available light	110" × 42" or 120" × 24"	¼"	3¼ lb.
Wired Dewdrop ...	About 75% of the available light	100" × 42" or 120" × 24"	¼"	3¼ lb.
Transparent Types Georgian Polished Wired	About 85% of the available light	110" long × up to 36" wide (greater widths can be supplied specially)	¼"	3¼ lb.

* GLAZING SIZES: This type of glass is largely used for roof and factory lighting in patent glazing bars, either vertical or horizontal, and in suitable sizes specified by the trade.

USES

Wired Glass is used for roof lights, lantern lights, and vertical glazing in public buildings, warehouses, factories, workshops, etc., where maximum protection is needed against shocks and risk of spreading fire. The transparent types are valuable for partitioned offices, counting houses, stores, etc., where clear view is desired, together with protection against breakage, fire and burglary.

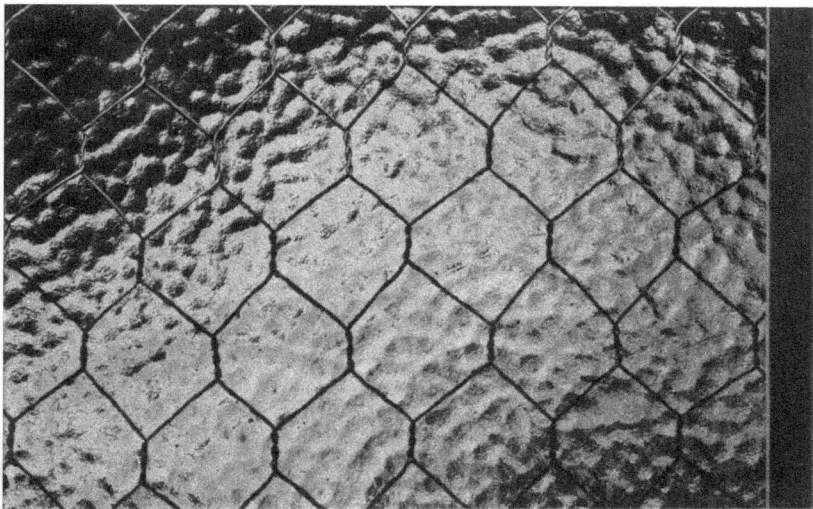

Wired Cast.   A rough cast double rolled glass with a $\frac{7}{8}''$ hexagonal mesh wire reinforcement.

Georgian Wired Rough Cast.   A rough cast double rolled glass, reinforced with fine $\frac{1}{2}''$ square mesh wire electrically welded at intersections.

Wired Arctic. *A figured rolled glass with a $\frac{7}{8}''$ hexagonal mesh wire reinforcement.*

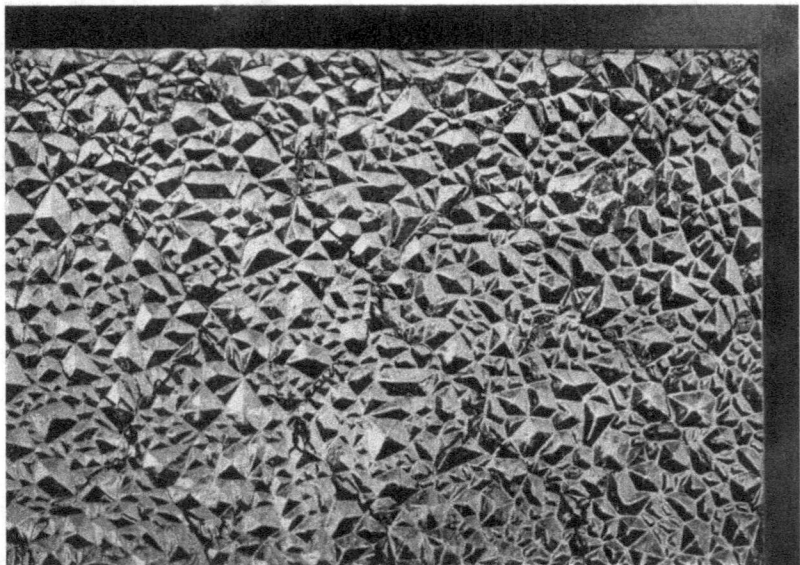

Wired Dewdrop. *A figured rolled glass, with a $\frac{7}{8}''$ hexagonal wire mesh reinforcement.*

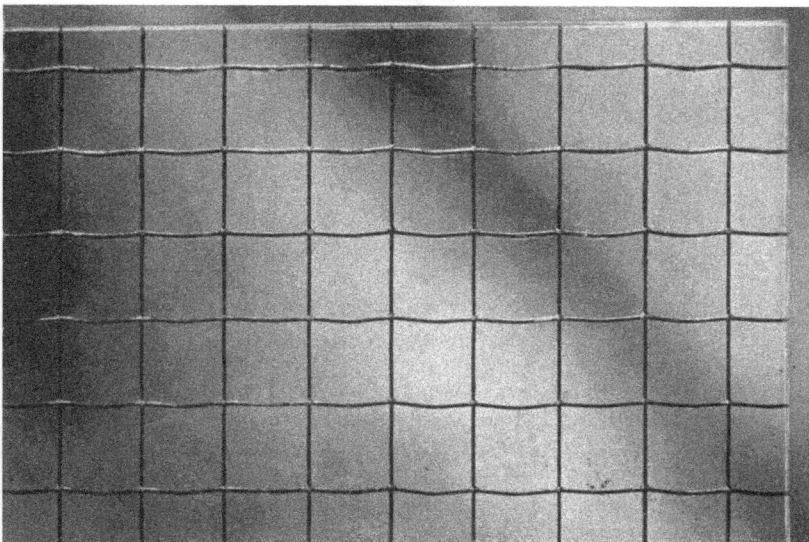

Georgian Polished Wired.   *Glass with a polished plate finish, reinforced with fine ½" square mesh wire electrically welded at intersections.*

FIRE-RESISTING GLAZING

The types of Wired Glass described on page 48 have been approved as a fire-resisting material when glazed in panes not exceeding two feet either way and secured with fire-resisting materials.

Local regulations vary in different parts of the country.

"ARMOURPLATE"
AND
TOUGHENED GLASSES

THE MANUFACTURE OF "ARMOURPLATE" AND TOUGHENED GLASS

In this process, Plate or Sheet Glass is suspended vertically in an electric furnace until the softening point is reached but the shape is retained. Then the glass is suddenly cooled by air directed to both surfaces of the glass which become solid, whilst the core is still hot and soft. As the core cools, it contracts, the outside layers being in compression, which makes the glass mechanically stronger. As the success of the process depends on having substantial outer layers in compression, it is not possible to toughen glass satisfactorily in thicknesses of less than 3/16".

Not only is toughened glass much stronger than ordinary glass of the same thickness, but when it is broken it breaks differently, disintegrating into small fragments.

Properties

Protection: "Armourplate" Glass if broken disintegrates into innumerable small pieces, not sharp enough to cause serious injury.

Resistance to Impact: When simply supported at the ends or along the edges, the resistance is increased to about seven times that of ordinary Plate Glass of equal thickness.

Resistance to Pressure: Transverse tests on sheets simply supported show "Armourplate" Glass to be four times as strong as ordinary Plate Glass of equal thickness, e.g. when a load is applied without shock to the centre of the surface, the breaking load for $\frac{1}{4}$" "Armourplate" Glass, size $45" \times 10"$ is 230 to 250 lbs.,

Demonstration of "Armourplate" Glass under load.

whereas for ordinary Polished Plate Glass of the same size and thickness, the breaking load is 50 lbs.

Resistance to Blast Pressure: Official tests have proved that "Armourplate" Glass is highly resistant to blast pressure. (Ref. A.R.P. Handbook No. 5, Structural Defence.)

Resistance to Wave Shock: Tests under conditions reproducing the effect of wave shock show that "Armourplate" Glass $\frac{1}{2}''$ thick will withstand pressure at least four times as great as that required to break ordinary Polished Plate Glass $1''$ thick.

Resistance to heat and sudden changes in temperature: Thermal tests show that "Armourplate" Glass offers great resistance to severe and sudden temperature changes. Provided that the heat is uniformly distributed "Armourplate" Glass will withstand temperatures up to 300° C. on one surface, with the other surface exposed to ordinary atmospheric temperature. It has also been tested to minus 70° C. without its quality being affected.

Qualities

"Armourplate" Glass retains the qualities of ordinary Polished Plate Glass, namely, transparency and brilliant lustre. It does not discolour under any conditions. Its expansion when heated is the same as that of ordinary Plate Glass.

Each piece of "Armourplate" Glass is indelibly branded "Armourplate" as a guarantee that it has been subjected to the special toughening process and has passed standard tests.

53

Sizes and Thicknesses

Thickness	Length	Width	
$\frac{3}{16}''$ i.e. 4.8/5.5 mm.	51"	25"	Sizes over 3ft. super should be as near 5.5 mm. as possible
$\frac{1}{4}''$	70"	52"	Sizes over 9ft. super to be in 17–20/64in.
$\frac{5}{16}'' - \frac{3}{8}''$	82"	70"	Also strips 9in. to 18in. wide × 11oin. long
$\frac{1}{2}''$	82"	70"	
$\frac{5}{8}''$	70"	52"	
$\frac{3}{4}''$ $\frac{7}{8}''$ $1''$ $1\frac{1}{4}''$			Can be supplied in sizes up to 8 sq. ft. Larger sizes in these thicknesses should be submitted to the manufacturers for consideration

"Armourplate" Glass can be supplied in most shapes if not too irregular. Any unusual shapes should be submitted to the manufacturers for consideration.

TOUGHENED GLASSES

The toughening process can be applied to certain other forms of glass, but the extent to which the strength of the glass and its resistance to temperature changes can be increased depends upon the type of glass used.

Embossed, sandblasted, painted and fired Toughened Glass can be supplied, but details should be submitted to the manufacturers for consideration.

Types

Toughened Black Glass, Toughened Rough Cast Double Rolled, Toughened Figured and Tinted Cathedral Glasses, Toughened Stippolyte and Selenium Glass, Toughened Tinted Polished Plate.

General

Any work on "Armourplate" Glass or Toughened Glass, i.e., embossing, brilliant-cutting, sandblasting or drilling of holes, must be carried out before the glass is subjected to the special treatment, as it cannot be cut or worked afterwards.

Holes should not be near the edge of "Armourplate" Glass and when bevelled glass is required, not more than $\frac{1}{8}"$ of glass must be removed, so that $\frac{1}{4}"$ glass must be left $\frac{1}{8}"$ thick on the edges; thus $\frac{3}{8}"$ glass must be left $\frac{1}{4}"$ thick on the edges, and so on. Care is necessary in handling and fixing "Armourplate" Glass so as not to damage the edge of the sheets by chipping. The edge of "Armourplate" Glass and Toughened Glass is not stronger than the edge of ordinary glass, and wherever possible the edge should be protected.[1]

USES

"Armourplate" Glass is used for automatic cigarette or ticket machines; battery assembling tables; gas and electrical cooker doors (single and double glazed); drying tables in chemical plants; display signs suitable for hanging outside hotels, etc.; glasses for meters, electric or gas floodlighting; for hospital locker, trolley and table tops; screens; shelves; windows for mental hospitals; miners' cap lamps; rough-usage mirrors (silvered); machinery guards; porthole glasses; drawing office flat printing frames; road signs; vacuum pan sight glasses; furnace flue inspection doors; fire screens; fire blowers; frameless entrance doors.

Other Toughened Glasses: for electric fires (Tinted Cathedral); fish and chip range backs; motor pit lights; underwater lighting; trawler floodlights; gas radiators; hospital equipment; windows in mental hospitals; police cell windows; road signs; shop fronts; canteen hot closet tops, etc.

[1] "Armourplate" is the registered trade mark of Pilkington Brothers, Limited.

"VITROLITE"

THE MANUFACTURE OF "VITROLITE"

"Vitrolite" is a coloured opaque glass, the opacity being due to the presence of small crystals in the glass. This crystallisation, which is really devitrification, is brought about by the addition of fluorides. The fluorides generally employed are fluorspar, cryolite or sodium-silico-fluoride. When a glass containing these fluorides is melted, the resultant molten glass is clear just like any ordinary glass, but on cooling, crystallisation of the fluorides occurs, resulting in a mass of fine crystals packed together in a matrix of clear glass. By varying the concentration of the crystals various degrees of translucency are obtained.

Coloured opaque glasses can be melted either in pots or tanks, and sheets are rolled in the manner of ordinary Rolled Plate, except that great care is necessary in rolling to produce a good fire-finished surface. The separation of the crystals occurs during or after the rolling, and as the degree of opacity depends largely on the rate of cooling, careful control of the temperature of rolling, speed of cooling and temperature of the annealing kiln is essential to give consistent results.

In the process of manufacture one surface of "Vitrolite" is impressed with a pattern of narrow parallel ribs to provide a key for the mastic or other material with which the glass is fixed.

Thicknesses Coloured "Vitrolite": $\frac{5}{16}''$ and $\frac{7}{16}''$
White and Black "Vitrolite" only: $\frac{5}{16}''$, $\frac{7}{16}''$, $\frac{3}{4}''$, $1''$.

Approximate average weight per sq. ft.

				Coloured		White		Black	
				lb.	oz.	lb.	oz.	lb.	oz.
$\frac{5}{16}''$	...	...	...	4	2	4	2	4	3
$\frac{7}{16}''$	...	...	...	5	10	5	10	6	0
$\frac{3}{4}''$	...	...	...			9	7	9	6
$1''$	...	...	...			12	12	12	11

An exterior view of an office building in which Black "Vitrolite" is used as a surface material, with Glass Bricks incorporated on the ground floor, and to admit light to the staircase. These are the offices of H. Wiggin & Co. Ltd., Thornliebank. Architect: M. Beckett.

Sizes

(1) Maximum Manufacturing Sizes:

$\frac{5}{16}$" and $\frac{7}{16}$"   125/140" long × 50/60" wide

$\frac{3}{4}$" and 1"   100/140" long × 50/55" wide

(2) "Vitrolite" can be supplied in stock sheets and cut sizes, but ashlar sizes are recommended and are less costly to fit.

(3) Ashlar Sizes:

$\frac{5}{16}$" "Vitrolite" is supplied, with edges true ground, in the following standard ashlar sizes:

10" × 15"	12" × 18"
15" × 15"	14" × 21"

(4) For special sizes, consult the manufacturers.

Colours

"Vitrolite" is supplied in the following colours:

Black, White, Green, Eggshell, Wedgwood, Turquoise, Pearl Grey, Green Agate, Ivory, Primrose, Shell Pink, Royal Blue Agate, Walnut Agate and Golden Agate, Tango and Cadmium Yellow.

The surface of "Vitrolite" is unaffected by water, soap, grease, damp, steam, all ordinary stains and acids except hydrofluoric.[1]

USES

"Vitrolite" is a wall lining material for hotels, schools, restaurants, dairies, bathrooms, lavatories, changing-rooms, operating theatres, hairdressing saloons, kitchens, etc. It can be used for corridor linings, shop-fronts, facias, counters, bars, table-tops, and for all kinds of display purposes, and for the exterior facing of buildings.

GLASS DOMES

THE MANUFACTURE OF GLASS DOMES

They are formed by bending a flat piece of glass over a mould. They are usually made in $\frac{3}{8}''$ Rough Cast Glass and in diameters rising by steps of 2″, from 18″ to 72″.

Standard Sizes

From 18″ diameter to 72″ diameter, with proportionate depths of from 2″ to 10″. Larger sizes are available but have to be made specially.

Four types of standard fittings can be supplied for fixing:

            (1) Type A (Light)
            (2) Type B (Standard)
            (3) Type C (Draught-proof for concrete curb)
            (4) Type C (Draught-proof for wood curb)

[1] "Vitrolite" is the registered trade mark of Pilkington Brothers Limited.

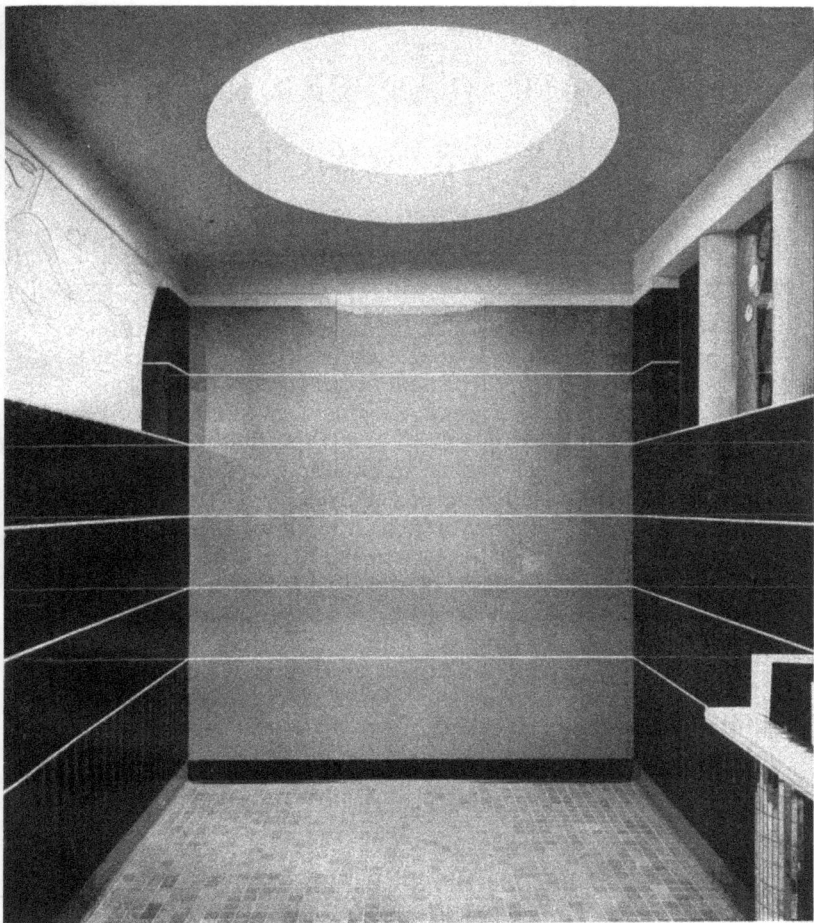

Lobby illuminated by Rough Cast Glass Dome

Domes can be supplied in Polished Plate if required, and it is also possible to introduce certain forms of decorative treatment.

USES Glass Domes are used to illuminate corridors and flat-roofed buildings.

INSULIGHT GLASS BRICKS AND "ARMOURLIGHT" TOUGHENED LENSES

Glass Bricks are made of soda lime glass, and the mixture of ingredients or "frit" is fed continuously into the furnace. Issuing from the furnace, the molten glass is cut off into "gobs" by automatic shears. Each gob, which contains the correct amount of glass to form one-half of a brick, falls into a mould on a rotary machine, and an automatic plunger presses the glass in the moulds to the correct shape.

The half bricks are then passed through an annealing lehr, after which pairs are sealed together to make a complete unit. In order to give the edges of the bricks a surface to which mortar will adhere, they are painted with a special paint, and sanded.

Insulight Glass Bricks can be laid like ordinary bricks. Other advantages are: thermal insulation; reduced condensation; little maintenance required; sound insulation; hygienic surface; fire resistance (classified Grade D under the British Standard Definitions No. 476, as a fire-resisting building material).

Fixing

Insulight Glass Bricks are non-load-bearing units which will carry their own weight with a wide safety factor up to any practical height, but because of wind pressures and other stresses it is necessary to put an intermediate support in panels over 20 ft. high or 120 ft. super. Very wide panels require an expansion joint every 20 ft.

Mortar: A fairly dry and fatty mortar is advisable as the Glass Bricks are non-absorbent. The best mix has been found to be 4 parts (by volume) sand, 1 part Portland Cement and 1 part

(1) **P.B.1:** $8'' \times 4\frac{7}{8}'' \times 3\frac{7}{8}''$. *Surface pattern of $\frac{1}{4}''$ convex ribs carried vertically on both exterior faces and horizontally on both interior faces. Approximate weight, 4 lbs. 5 ozs.*

(2) **P.B.2:** $5\frac{3}{4}'' \times 5\frac{3}{4}'' \times 3\frac{7}{8}$. *Surface pattern of $\frac{1}{2}''$ convex ribs carried vertically on both exterior faces and horizontally on both interior faces. Approximate weight, 3 lbs. 11 ozs.*

(3) **P.B.2 Corner Brick:** *Surface pattern of $\frac{1}{2}''$ convex ribs carried vertically on both exterior faces and horizontally on both interior faces. Approximate weight, 3 lbs. 10 ozs.*

(4) **P.B.3:** $7\frac{3}{4}'' \times 7\frac{3}{4}'' \times 3\frac{7}{8}''$. *Surface pattern of $1\frac{1}{4}''$ concave ribs carried on both interior faces, running vertically on one face and horizontally on the other. Both exterior faces are smooth. Approximate weight, 6 lbs.*

(5) **P.B.3 Corner Brick:** *Surface pattern of $1\frac{1}{4}''$ concave ribs carried on both interior faces, running vertically on one face and horizontally on the other. Both exterior faces are smooth. Approximate weight, 7 lbs. 10 ozs.*

(6) **P.B.32:** $7\frac{3}{4}'' \times 7\frac{3}{4}'' \times 3\frac{7}{8}''$. *Surface pattern of $\frac{1}{2}''$ convex ribs carried vertically on both exterior faces and horizontally on both interior faces. Approximate weight, 6 lbs.*

(7) **P.B. 32 Corner Brick:** *Surface pattern of $\frac{1}{2}''$ convex ribs carried vertically on both exterior faces and horizontally on both interior faces. Approximate weight, 7 lbs. 11 ozs.*

61

Interior view of the three panels of Glass Bricks which are incorporated in the whole length of a staircase, and which provide privacy with high light transmission. (For the external use of Glass Bricks, see photograph illustrating the use of Black "Vitrolite" on page 57.) This staircase is in the offices of H. Wiggin & Co. Ltd., of Thornliebank.   *Architect: M. Beckett.*

slaked lime putty, mixed fairly dry. The sand should be clean builders' sand free from gravel. Slaked lime putty has been found to be more fatty and therefore preferable to hydrated lime (i.e. dry lime or bag lime).

Pointing: The face of the joints may be struck back and smoothed during the erection, or they may be raked out and later pointed up with Snowcrete, Colourcrete or similar materials. A "keyed" joint formed with a curved jointing tool is the finish mostly used. An alternative method of adding "colour" to a panel is to paint the joints with a good oil paint or aluminium paint after the mortar is quite dry. This method should only be used on internal panels.

Reinforcement: Reinforcing strips should be built in every third to every fifth course according to the size and position of the panel. The ends should pass through the clearance joint and be built into or secured to the main structure. "Exmet" $2\frac{1}{2}''$ wide No. 20 gauge expanded metal has been found most suitable for this purpose.

Clearance Joints: The top and both vertical edges of every panel must be built free of the main structure—except for the reinforcement—to avoid risk of settlement, load or expansion strains affecting the panel. A $\frac{1}{2}''$ clearance is advisable and this should be kept free of any spillings of mortar and be filled with a non-hardening material. Griponex mastic has been found most suitable for filling the clearance joints on exterior panels, and Gripon 10.10.N. for interior panels. There are, however, many non-hardening bituminous and other compositions on the market. Wherever possible the top and ends should be built into a "recess" to provide stability not otherwise obtainable owing to the clearance joints. This recess should be $4\frac{1}{4}''$ wide by $1''$ deep, allowing $\frac{1}{2}''$ clearance and $\frac{1}{2}''$ cover over the face of the Glass Brick, with $3/16''$ play on either face, and this also should be pointed with a non-hardening composition. Only small internal panels should be built into a "rebated" joint, i.e. recessed $1''$ on one face only, as a plain "butt" with no cover on the face to provide stability is not generally advised, and schemes requiring this type of

joint should be submitted to the manufacturers before allowing the job to be continued.

USES
Eminently suitable for external or internal walls for staircase
lights, panels, bays, partitions; for flat or curved surfaces;
for any position where a diffused all-over light is required,
or light combined with privacy.

MANUFACTURE OF LENSES, AND TOUGHENING PROCESS
Pressed glass used in glass and concrete construction is various-
ly described as "Tiles", "Lenses" or "Pavement Lights"—
all these being made by pressing. In this process the tile is
formed in a cast-iron mould which is usually built in two
halves hinged together and mounted, with a base plate, on a
press, operated by a lever-motion or by a compressed air
cylinder. The press carries a vertical spindle on which is
mounted the cast-iron plunger. The mould and plunger are
shaped and patterned, a high quality cast iron being used,
so that they will have a good surface finish.

Process: Molten glass at a working temperature of approximately 1,100° C. is taken from a tank-furnace by means
of a long-handled steel ladle, in which the glass is carried to
the press and poured slowly into the mould. When the right
quantity of molten glass has been poured, the operator cuts
through the stream of glass with a pair of hand shears. The
press is then put into operation and the plunger brought
down into the mould with sufficient pressure to force the glass
to fill all the space between the mould and plunger. The plunger remains in the mould until the glass has set sufficiently to
keep its shape; it is then withdrawn, and a jet of cooling air is
directed on to the tile, which, when sufficiently rigid, is
transferred to an annealing lehr, where internal strains are removed, and the glass cools slowly to atmospheric temperature.

As the tiles are taken from the delivery end of the lehr they
are carefully examined for shape, size and quality. Batches
are regularly examined by inspection in polarised light to
check the degree of annealing.

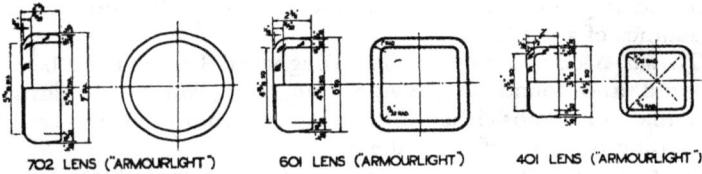

702 LENS ("ARMOURLIGHT")   601 LENS ("ARMOURLIGHT")   401 LENS ("ARMOURLIGHT")

Toughening: Certain types of glass tiles can be toughened during manufacture by a process very similar in principle to that used for the toughening of flat glass. The tile is put on a toughening machine, where it undergoes a uniform cooling by compressed air. From the toughening process it passes through two routine tests, firstly by transferring it from an oven at a temperature of 120° C. to a bath of cold water, and secondly by transferring it quickly from atmospheric temperature into an oven at a temperature of 500° C. where it remains for 5 minutes. The first test ensures that the tile is properly toughened and the second eliminates any tiles with weaknesses in the tension zone.

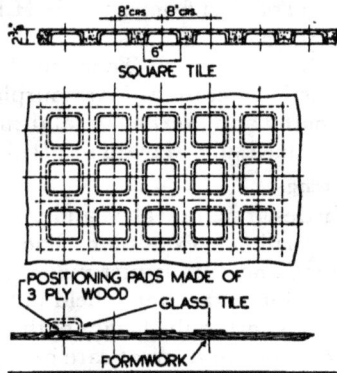

Diagrams showing types of lenses

Properties

"Armourlight" Toughened Lenses have an exceedingly high resistance to mechanical loads and severe thermal conditions, and will withstand an impact test approximately 10 times as great as an annealed tile of similar shape and of twice the thickness. Toughened Lenses may be set direct in concrete, without risks of cracks developing as a result of the stresses induced by movement of the concrete.

A double glazed unit of "Armourlight" Toughened Lenses

65

provided protection against a 4.5 revolver bullet fired at a distance of 5 yards.

Thermal Shock: "Armourlight" Toughened Lenses have been tested and found to provide complete protection against incendiary bombs either of the kilo-electron or thermite type, burning on the surface of the lens.

Blast: A panel 4′ 6″ × 3′ of Type T.702 Lenses withstood the effect of blast from a 500 lb. H.E. bomb detonated at a distance of 50 feet.

"Armourlight" Toughened Lenses have the same characteristic fracture as "Armourplate" Glass, and if broken, they expand and exert a lateral thrust on the concrete in which they are fixed, and remain in position.

Fixing

In designing the glass and concrete structures, allowance must be made for expansion. The "formwork" for casting the concrete and tiles is first made, and the tiles are then placed in position and suitably held until the concreting is finished. The slab is cast to the height of the reinforcing rods, which are then set in position, great care being taken that they are clear of the tiles: the casting is then proceeded with and completed.

Great care should also be taken to see that the sides of the tiles making contact with the concrete are quite clean and free from grease, dirt or foreign matter of any kind to ensure complete adhesion of the concrete to the glass.

A suitable mixture for this class of work is a 3 : 2 : 1 "mix" consisting of three parts $\frac{1}{4}$″ down to $\frac{1}{8}$″ granite free from dust, two parts washed pit or river sand (not more than 20% passing a 50-mesh sieve), and one part cement. Use clean water in mixing the concrete, which should be of a sufficiently plastic consistency to permit even flow of the material into all parts of the formwork and around reinforcement. The concrete must be well "punned" so as to be solid and free from air bubbles.

The top of the tiles can be cleaned when the surface of the concrete is smoothed over, which is best done a few hours after the initial "set."

Water curing in the initial stages is strongly advised. This

66

An installation in which Type T.601 Toughened Glass Lenses have been used, at the Five Ways Hotel, Nottingham. As illustrated, Toughened Roof Lenses ensure a clean, neat job with excellent light transmission, and risk of breakage is eliminated.

An example of the external use of Type T.601 "Armourlight" Glass Lenses, at the Five Ways Hotel, Nottingham.

is best carried out by covering with hessian or other suitable material, which can be kept continuously damp. This is done to ensure slow maturing of the concrete and should be carried out for at least four days.

USES

"Armourlight" Toughened Lenses are used for roof lights, and also vertical lights in the form of pre-cast slabs.

The patented process of toughening moulded and blown glass is a development of the "Armourplate" process. "Armourlight" is the trade name given to toughened blown and moulded glasses, which include bulkhead glasses, well glasses, miners' lamp glasses, etc., and toughened lenses. Whereas the degree of toughening in "Armourplate" is constant, owing to the even substance of Plate Glass, the degree of toughening in "Armourlight" varies because of the difference which occurs in the shape to be toughened and the variation in substance. Accordingly, the resistance to impact and thermal shock also varies in different articles.

SPECIAL GLASSES

"VITA" GLASS

"Vita" Glass permanently admits the passage of short-wave ultra-violet radiation. This radiation has therapeutic properties, from 3,200 A.U.* to the limit of the solar spectrum. Ordinary glass is opaque to such radiation.

Thicknesses and Weights

	Nominal Thickness of Substance	Thickness Tolerance	Weights per sq. ft.
Plate "Vita"	$\frac{5}{32}''$	0.14″–0.19″	1 lb. 13 oz.–2 lb. 7 oz.
Sheet "Vita"	18–20 oz. 26 oz.	0.08″–0.11″ 0.12″–0.13″	1 lb. 1 oz.–1 lb. 7 oz. 1 lb. 9 oz.–1 lb. 12 oz.
Cathedral "Vita" ...	$\frac{1}{8}''$	—	about 1½ lb.
Wired Georgian "Vita"	$\frac{3}{16}''$	0.19″–0.21″	2 lb. 7 oz.–2 lb. 13 oz.

It is used in modern hospital construction (ordinary wards or specially constructed balconies where heliotherapy is practised), in sanatoria, nurseries, schools, sun-lounges, etc., and also for the glazing of essential rooms.

As ultra-violet light is not only emitted by the sun's rays but also reflected from blue sky and white clouds, "Vita" glass may appropriately be used in windows facing north.

NON-ACTINIC

A Rolled Glass of a soft greenish tint, which is opaque to ultra-violet light. Its texture is that of ordinary tinted Double Rolled Cathedral Glasses. To secure the fullest advantage from

* Angstrom Unit equals one ten-millionth of a millimetre.

the use of this glass, provision should be made for the absorbed heat to be dissipated by natural upward movement of air against the interior surface of the glass. Normally this would be provided for by ordinary roof and wall ventilators such as are fitted in all tropical buildings.

Nominal Thicknesses: $\frac{1}{8}''$, $\frac{3}{16}''$, $\frac{1}{4}''$.

Maximum Manufacturing Sizes: Up to $110'' \times 24''$, $100'' \times 36''$, $90'' \times 42''$.

This glass, being opaque to ultra-violet light, greatly reduces the fading action of daylight and is therefore useful for glazing where there is no objection to the colour of the light transmitted. "Non-Actinic" also absorbs a proportion of the sun's heat. In sub-tropical and tropical climates, Non-Actinic Glass is used for ample but restful illumination, with safety against eye-strain and sun-stroke.

CALOREX

Made in clear Sheet where clear vision is needed, and in rolled form (Calorex Cast) where diffusion is desirable. It is of greenish-blue tint. Used chiefly for roof glazing.

Qualities: Calorex Sheet is made in one quality in two thicknesses, 21 oz. and 32 oz. per square foot, and in sizes $40/56'' \times 32/36''$.

Calorex Cast is made in three thicknesses, $\frac{1}{8}''$, $\frac{3}{16}''$, and $\frac{1}{4}''$, and in sizes up to $90'' \times 36''$.

Calorex is also obtainable in the form of ground and polished plates in sizes up to $60'' \times 36''$ in a thickness of $\frac{1}{4}''$.

Light Transmission: Approximately 60% of the available light, this figure being independent of the thickness of the glass, since the mixture is adjusted to give a constant transmission in the nominal thickness.

Special Uses

Calorex absorbs about 80% of the sun's heat irrespective of the thicknesses indicated above, while at the same time transmitting a considerable portion of the visible radiation.

70

It is, therefore, particularly suitable for such buildings as factories, warehouses and garages, where extremes of temperature are undesirable. Its principal use is for glazing summer houses, verandahs, sports courts, abattoirs, and hospitals; it is extensively used in tropical countries. It is a matter of interest that Calorex has been shown to have qualities that are deterrent to flies and insects.

ANTI-FLY GLASS

Tests have been carried out to discover the effect of different colours on flies and other insects. In these tests the flies were allowed to live under any colour they chose, and it was found that in general they preferred white light to coloured light and that red was the best deterrent, the other colours following in the order of the spectrum. It would be impracticable to use red glass owing to the loss of light. Yellow glass is almost equally effective. Although the use of yellow glass is not an absolute preventive, it is an effective deterrent, and is of value for use in connection with the storage of food. Yellow glass is not advisable in buildings where people are constantly employed, because yellow light is injurious to the eyes and general health. Apart from this effect of colour, it was found that although in warm weather coloured glasses acted as deterrents to flies, in cold weather they would rather be in a warm place, whatever the colour of the glass.

A special amber-tinted glass which is an effective deterrent and valuable for use in buildings connected with the storage of food, is known as Anti-Fly Glass.

NEW CROWN

Made in one quality and thickness only, nominally 18 oz.

Light Transmission: 90% of the available light.

Manufacturing sizes: 18″ × 12″.

Special Uses

For re-glazing windows in old houses if it is desirable to maintain the original appearance.

ANTIQUE REAMY

A hand-blown glass of uneven thickness, both surfaces of which are textured. It is made chiefly for stained glass windows, since its surface simulates mediaeval glass used for that purpose. Other types of antique glass are made for special purposes.

Sizes and Thicknesses: The average size of the sheets obtainable is 24″ × 15″. The thickness varies in each sheet from $\frac{1}{8}$″ to $\frac{3}{16}$″ approximately.

This glass is used for church windows and similar glazing, or glazing of a decorative character.

PRISMATIC GLASS

A translucent Rolled Glass, one surface of which consists of parallel prisms arranged in such a manner that light passing through the glass is refracted in a certain direction according to the incidence of the light and the angle of the prism; the other surface is smooth.

Thickness and Weight: Nominal thickness: $\frac{1}{4}$″. Approximate weight: 3 lb. per sq. ft.

Light Transmission: 50% to 90% of the available light, according to the direction in which the transmission is measured.

Quality: Made in one quality only.

Manufacturing sizes: 60″ high × 100″ wide, the prisms running with the width of the sheet.

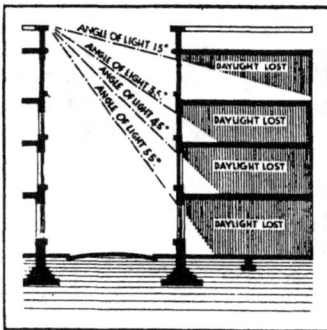

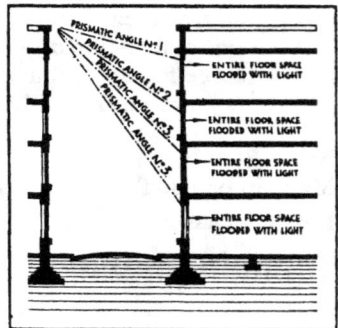

NOTE: The glass should always be fixed with the prisms running horizontally and on the inside of the window.

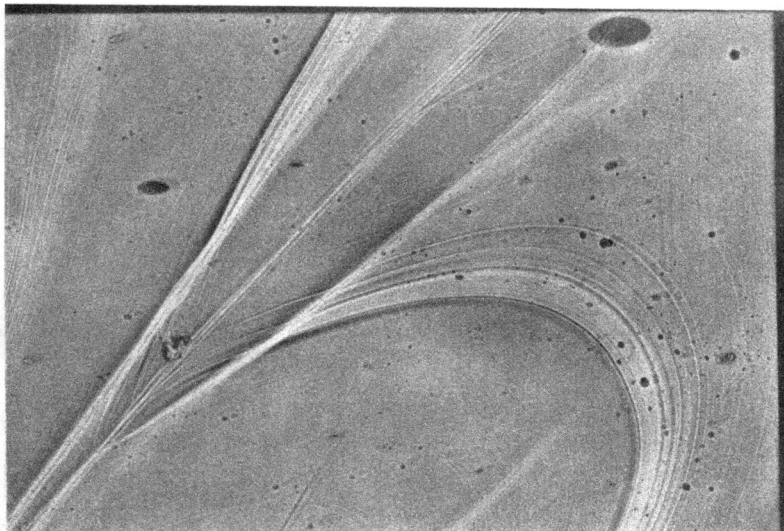

Antique Reamy

Prismatic, Angle No. 2

Prismatic Glass is used for glazing windows which are overshadowed by neighbouring buildings; to transmit light into dark places; and to ensure maximum use of available daylight.

Angle No. 1: For situations where the angle of the light's incidence taken from the horizontal is up to 30°.

Angle No. 2: For situations where the angle of the light's incidence taken from the horizontal is between 30° and 40°.

Angle No. 3: For situations where the angle of the light's incidence taken from the horizontal is over 40°.

MAXIMUM GLASS

A Rolled Glass, one surface of which consists of parallel prisms which are so designed that light received from a relatively high elevation is refracted horizontally after passing through the glass. The outer surface is fluted at right angles to the direction of the prisms in order to give suitable lateral diffusion.

Qualities: Made in two forms. Maximum Angle A, for use when the angle of elevation of the obstruction is not more than about 25°–30°; and Maximum Angle BD, for use when the angle of elevation of the obstruction exceeds 40°.

Light Transmission: The light transmission varies according to the side on which the light is incident, and according to the angle, but may be taken as being between 65% and 70% of the available light.

Thickness and Weight: Nominal thickness is $\frac{3}{16}$″. Approximate weight per square foot $2\frac{1}{4}$–$2\frac{1}{2}$ lb.

Manufacturing sizes: 100″ × 40″, the prisms running either with the length or width of the sheet.

Special Uses

For glazing windows which are over-shadowed by neighbouring buildings, such as basements or offices whose windows open into an area or well.

The effect of this glass is to refract the light of the sky horizontally into the room.

Maximum Angle A

FLASHED GLASSES

Blown Sheet Glass with a thin layer of opal glass flashed during manufacture upon one surface. White opal may be flashed on to Tinted Sheet, to furnish a range of colours. Usual substances or thicknesses: 15–18 oz. and 21–24 oz. Manufactured in one quality only.

Flashed glasses are used for artificial lighting, laylights, lighting fittings, etc.

NEUTRAL TINTED

A dark grey transparent Polished Plate Glass which does not appreciably change the hue of light passing through it. Used as an anti-glare glass and in ambulance windows, etc., for privacy.

BROAD REEDED

A Rolled Glass with a pattern which consists of wide, concave cylindrical lenses imprinted on one side. It does not offer a high degree of obscuration.

75

It is manufactured in one quality only, in white glass.

Special Uses

In windows, doors, partitions, screens, laylights, lighting fittings. When silvered on one side, it is suitable for bathroom walls, fire-places, surrounds, etc.

CROSS REEDED

A Rolled Glass similar to Reeded except that the pattern is impressed on both sides, the cylindrical lenses being at right angles. The double set of lenses, which in this case are convex instead of concave, cause increased obscuration.

CHEVRON REEDED

Rolled Glass with the reeds running across the sheet in chevron formation. This glass affords complete privacy with a high transmission of light.

Quality: One quality only, manufactured in white glass.

Special Uses

In windows, doors, screens, etc.

REEDED

A Rolled Glass, similar to Broad Reeded, except that the concave cylindrical lenses are not so wide.

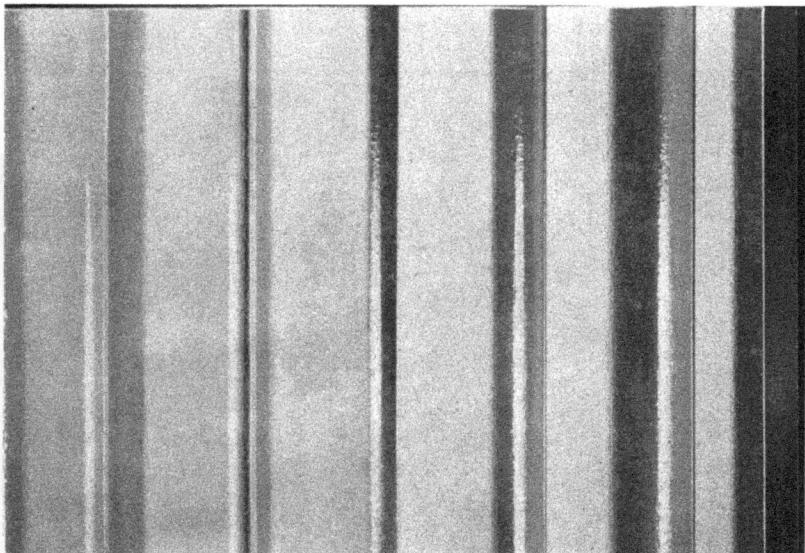

Broad Reeded

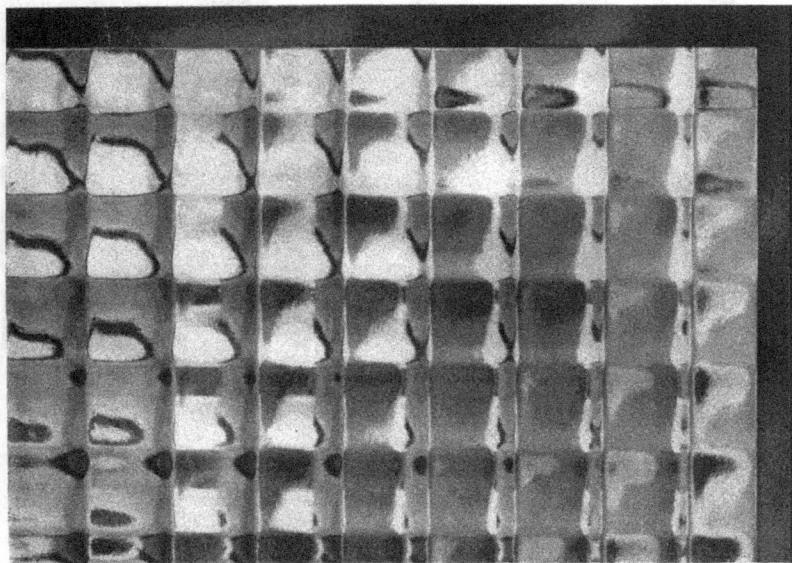

Cross Reeded

Chevron Reeded.

Installation of Broad Reeded at the Housing Centre, Suffolk Street, London, S.W.1.

SOME DECORATIVE TREATMENTS FOR GLASS

There are three main processes by which the surface of glass can be worked:

Embossing (or acid etching): Hydrofluoric acid has the property of dissalving glass. When it is applied to, say, clear Polished Plate Glass it eats away the surface, but leaves it comparatively clear. By neutralising it, through the addition of an alkali such as ammonia, a dense, white, frosted appearance is obtained. This combination of acid and alkali goes by the name of White Acid, and is the principal agent in the compound process known as French, or Triple Embossing. The variations which constitute Acid Embossing are merely variations of tone.

Stippling: By strewing grains of mica evenly over the Plate before flooding with the acid, a texture which is known as acid stipple can be given to the glass. The coarseness or fineness of the stipple is determined by the size of the grains of mica.

Brilliant Cutting: This is one of the oldest methods employed in the decoration of glass. It is carried out by means of a revolving stone, it being possible to obtain a square cut, or mitre cut (an incised "V" line), or the edge cut (an unequal angle "V" line), "punties" (circular-shaped cuts), or "hollows" (elongated punties). When the line has been cut by the stone it is usually polished by means of a revolving wood wheel and finished off with a brush. An unpolished brilliant cut line, however, on say a mirror, shows up with much greater precision than a polished line.

Other decorative processes are:

Sandblasting: This is an operation where the whole or part of the surface of a piece of glass can be obscured, by forcing sand

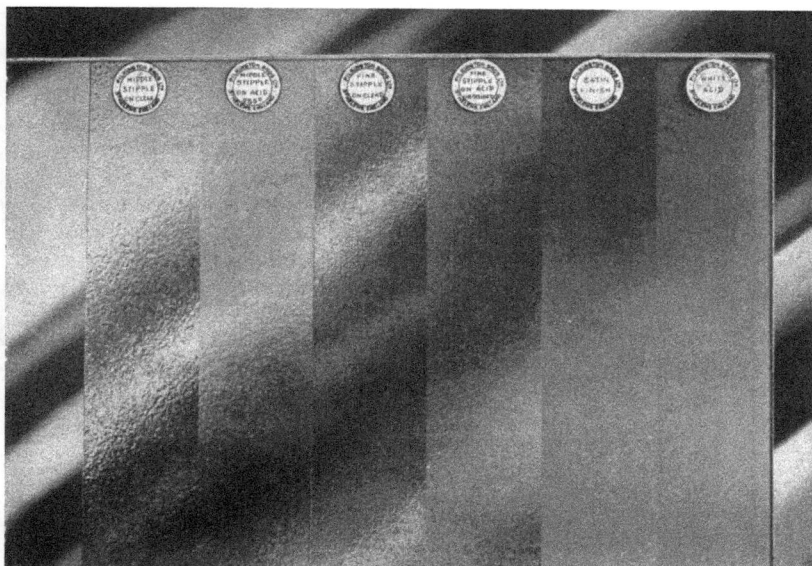

Surface treatments, standard acid obscuration (Plate Glass).

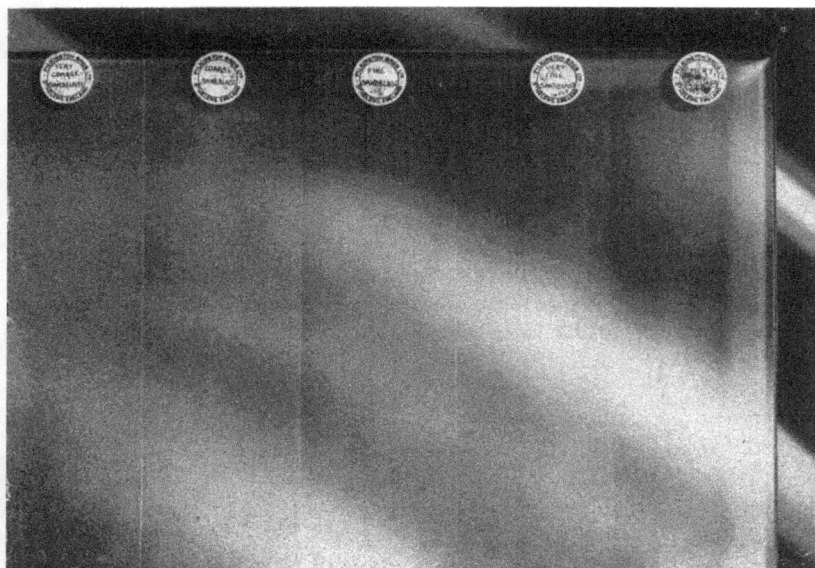

Surface treatments, standard emery and sandblast obscurations (Plate Glass).

under pressure, through the nozzle of a "gun". The sand, forced out at an enormous pressure from the gun, destroys the highly polished surface of the glass, which then appears slate-grey in tone. An alteration to the size of the nozzle of the gun, the quality of the sand, and the air pressure of the machine, results in work fine enough for the decoration of a wine-glass. The "resist" used to mask that portion of the design which is not to be sandblasted, is made from paper, one side of which has been treated with a special preparation. Various finishes are possible:

(a) *deep sandblast:* on thick Plate Glass this has the effect of modelling, or if on the back of the glass, an effect of raised relief.

(b) *medium sandblast:*

(c) *light sandblast:* a flat obscured effect.

(d) *peppering:* sandblast very lightly applied so that the polished surface is not entirely removed. By very finely sandblasting a plate, it can be almost as smoothly and delicately obscured as with acids.

(e) *shaded sandblast:* particularly effective on Black Polished Plate Glass, or Black "Vitrolite".

Artificial light directed from the edge through the thickness of a plate of glass, impinges on any sandblasted incisions in the surface of that plate, and the sandblasted lines or pattern take up the illumination, and glow in apparent independence of the source of light.

Silvering: Formerly, mirrors were coated with an amalgam of tin and mercury. This process has been superseded by the chemical deposition of silver. Tinted mirrors can be made by silvering coloured Polished Plate. "Stripping" is the term used when silvered backing is completely removed from a mirror, and "brushing-out" is the expression used to denote the part-removal of the silvering. Gold, as well as silver and some other metals, may be deposited electrolitically, on glass, with many beautiful results. Platinum painting is one of the processes, the metal being applied in the form of paint, to the surface of the glass, which is then fired in a kiln.

BRITISH STANDARDS INSTITUTION SPECIFICATIONS

The British Standards Institution have published standard specifications as follows:—

No. 952—1941: British Standards for glass for glazing including definitions and terminology of work on glass.

No. 973—1941: British Standard Code of Practice for the glazing and fixing of glass for buildings.

No. 990—1941: War Emergency British Standards Institution, metal windows and doors.

BIBLIOGRAPHY

A Key to Modern Architecture, *by F. R. S. Yorke, A.R.I.B.A., and Colin T. Penn, A.R.I.B.A.*

A Text-Book of Glass Technology, *by A. Cousen and F. W. Hodkin.*

An Introduction to Modern Architecture, *by J. M. Richards (Penguin Books Ltd.).*

Architecture, *by Christian Barman (Benn's 6d. Library).*

Architecture, *by W. R. Lethaby (Home University Library).*

Balbus, or the Future of Architecture, *by Christian Barman.*

Department of Scientific and Industrial Research, Technical Papers:
  Penetration of Daylight and Sunlight into Buildings, 1932.
  Surface Brightness of Diffusing Glassware for Illumination, 1926, *C. J. W. Grieveson*
  Terminology of Illumination and Vision, 1935
  The Natural Lighting of Picture Galleries, 1927.
  The Transmission Factor of Commercial Window Glasses, 1926, *A. K. Taylor and C. J. W. Grieveson.*
  The Transmission of Light through Window Glasses, 1936.

English Glass, *by W. A. Thorpe.*

Form in Civilisation, *by W. R. Lethaby.*

Glass and Glass Manufacture, *by Percival Marson, published by Sir Isaac Pitman & Sons Ltd.*

Glass in Architecture and Decoration, *by Raymond McGrath, B.Arch., A.R.I.B.A. and A. C. Frost, B.A.Cantab.,* with a section on the nature and properties of Glass, *by H. E. Beckett, B.Sc.*

Glass Manufacture, *by W. A. Rosenhain.*

Glass: the Miracle Maker, its history, technology and applications, *by C. J. Phillips (Pitman Publishing Corporation, New York and Chicago).*

Glassmaking in England, *by H. J. Powell.*

In the Nature of Materials: the Buildings of Frank Lloyd Wright, 1888-1941, *by Henry Russell Hitchcock* (Duell, Sloan and Pearce, New York).

Industrial Art Explained, *by John Gloag.*

Les Propriétiés physiques et la fusion du verre par *Bernard Long* (Paris, Dunod).

Machinery and Methods of Manufacture of Glass, *by Professor W. E. S. Turner.*

Modern Glass Practice, *by S. R. Scholes (Industrial Publications Inc., Chicago).*

New Sights of London, *by Hugh Casson, A.R.I.B.A.,* with an introduction *by John Gloag* (London Passenger Transport Board, 6*d.*).

Properties of Glass, *by G. W. Morey (American Chemical Soc. Monograph Reinhold Publishing Corporation, N.Y.).*

Smaller Retail Shops, *by Brian and Norman Westwood* (No. 2 in the Planning of Modern Building series.)

The Constitution of Glass, *by Professor W. E. S. Turner.*

The Modern Flat, *by F. R. S. Yorke, A.R.I.B.A., and Frederick Gibberd, A.I.A.A.*

The Modern House in England, *by F. R. S. Yorke, A.R.I.B.A.*

The Principles of Architectural Composition, *by Howard Robertson, F.R.I.B.A.*

The Works of Man, *by Lisle March Phillipps.*

Theory and Elements of Architecture, *by Robert Atkinson, F.R.I.B.A., and Hope Bagenal, A.R.I.B.A.*

Twentieth Century Houses, *by Raymond McGrath, B.Arch. A.R.I.B.A.*

100 Years of British Glass Making, *Chance Brothers and Co. Limited.*

1851 and the Crystal Palace, *by Christopher Hobhouse.*

A BRIEF GLOSSARY OF TERMS
CONNECTED WITH GLAZING

ACIDING   See Embossing.

ALLOWANCE   The difference between the glazing or glass size and the size of the rebated opening into which glass or a glazed frame is to be fitted.

ARRISSING   The process of removing the sharp edges of glass after it has been cut with a diamond.

BACK PUTTY   The putty making a seal between the glass and rebate after the glass has been pressed into position.

BAIT   An iron grille lowered into the molten glass and to which the glass adheres. The grille is gradually raised followed by a continuous ribbon of Sheet Glass.

BEAD   A strip of wood or metal attached to the inside or out-side of the frame to secure the glass. Metal or hard-wood beads are fixed by means of brass-cupped screws. Beads are invariably used when glazing with Plate Glass.

BEND   A pane of glass which has been bent to fit an opening which is curved on plan, in section, or both, is known as a bend. A pane of glass, such as a segment of a dome, which is bent in two directions, is referred to as a "double bend".

BEVEL   Bevelling is a hand, or machine process, whereby the surface edges of glass are ground, smoothed and polished, or ground and smoothed, or ground only, according to the type of bevel required.

BORROWED LIGHT   A window inside the building which transmits light obtained from external windows.

BRILLIANT CUTTING   A decorative process employed for cutting designs on glass whereby various types of cuts are made by bringing the plate to bear on a wheel of the required section, the cuts, unless desired otherwise, being subsequently smoothed and polished.

BULLION OR BULL'S EYE   The characteristic scar or lump in the centre of the crown disc formed during the transference of the glass from the blowing iron to the punty.

CAME   The metal strip used for leaded lights and copperlights. Steel-cored cames are sometimes used for leaded lights to obviate the need for saddle bars.

CASTING   The process of pouring molten glass and rolling it into a flat sheet.

85

GLOSSARY

COPPERLIGHTS	"Copperlights" or "copperlight glazing" is the generic term applied to the type of fire-retarding glazed lights built up, in the manner of leaded glazing, with electrically welded cames of copper.
CULLET	Broken glass.
CUT SIZES	Glass cut to size ready for glazing is referred to as squares, panes, or cut sizes.
DIAMOND	A quarry, or pane of glass, cut in the shape of a diamond.
DOME LIGHT	A roof-light in form of a dome, constructed either as a one-piece glass dome or in segments of bent glass framed in wood or metal.
DOUBLE EMBOSSING	An embossed surface of glass having three shades of obscuration.
DOUBLE GLAZING	Glazing in which, for purposes of insulation, the inner and outer panes are fixed in separate frames.
DRAWBAR	A block of refractory material submerged into the molten glass, and round either side of which the glass flows.
EDGE-WORK	The term applied to the hand, or machine work done to the cut edges of glass. The edges of the glass may be ground, smoothed and polished, or ground and smoothed, or ground only, according to the type of edge-work required.
EMBOSSING OR ACIDING	A process whereby the surface of glass is obscured by by treatment with hydrofluoric acid or its compounds.
FASCIA	A flat member in an entablature: a broad horizontal band forming one of the members of a cornice. In a shopfront the fascia is the band above the sash which usually bears the inscription.
FIRE-RETARDING GLAZING	(Originally termed Fire-resisting). Glazing, such as wired glass or copperlight, which complies with the regulations laid down by the London County Council, Firemen's Association, etc.
FIRE-FINISH	The term applied to glass which retains its natural fire-polished surface. With fire-finished glass there is always a certain degree of distortion of vision and reflection.
FIXED LIGHT	A window which is not made to open. In a range of windows the casements or other hinged frames are referred to as "opening lights".
FRENCH EMBOSSING	See Triple Embossing.

86

FRENCH WINDOWS	Folding casements, without fixed meeting-rails, which extend to the floor level, and are used as doors.
FRIT	The raw material from which glass is made after having been ground and mixed together.
GILDING	A process, employed largely for lettering and decorative work, whereby leaf metal, such as gold leaf, is applied to the surface of glass and coated with a protective medium.
GLAZING	A term applied to the securing of glass in a prepared opening.
GLAZING BAR	An intermediate member in a window unit.
GLAZING GROOVE	A groove made to receive the glass: used instead of a rebate, e.g. on the meeting-rail of the lower sash of a double-hung window.
GOB	Molten glass flowing from the tank in predetermined quantities for the making of individual glass articles.
GRAVE OR MODELLED SANDBLAST	Deep, as opposed to surface sandblasting.
GRINDING	A process whereby the surface of the glass is obscured by grinding with an abrasive. In general, grinding produces a finer surface than sandblasting, the fineness of the surface depending on the nature of the abrasive.
HEAD	The topmost horizontal member of a sash.
HOPPER WINDOW	A sash hinged at the bottom and usually fitted with side cheeks to prevent cross draughts in a room.
JAMB-LINING	The material used to line the door or window opening.
LANTERN LIGHTS	The glazed vertical or sloping sides of a skylight.
LAYLIGHT	A ceiling light for internal artificial lighting.
LEADED LIGHT	A panel made up of small pieces of glass in lead cames.
LEHR	A long heated tunnel consisting of gradually falling temperatures.
LIGHT	A window or similar construction for the admission of light, e.g. roof-light, pavement light, borrowed light, north light.
MARGINAL LIGHTS	Glazed frames at the sides of a door embodied in the door frames.
MASTIC	A permanently plastic material used for bedding glass for internal and external wall linings.
METALLISING	A process whereby metals other than silver are deposited on glass with or without a protective coating.

MIRRORS
Selected Polished Plate Glass silvered on one side.

OBSCURING
PROCESS
Processes such as sandblasting, grinding and aciding or embossing, employed to treat the surface of glass after manufacture whereby vision through the glass is obscured to a varying degree and the light-diffusing properties of the glass are increased.

PAINTING OR
STAINING AND
FIRING
A process whereby glass is first coated with a fusible pigment and subsequently fired. Staining is usually applied to more or less transparent colours and painting to dense or opaque colours.

PATENT GLAZING
The generic term covering all forms of glazing which rely for their efficiency upon some means of collecting and carrying away water by means of channels or grooves incorporated in the glazing bar, as distinct from putty and similar glazing.

POLISH
The term applied only to the surface of glass which has been ground, smoothed and polished, the object being to obtain clear, undistorted vision and reflection, as in the case of Polished Plate.

PUNT
A circular or oval depression in the surface of glass, made by brilliant cutting.

PUNTY
(*Crown Glass process*)
An iron rod to which Blown Glass was transferred, prior to spinning out into a disc.

QUARRY
A pane of glass, especially one cut to the shape of a lozenge or diamond and lead-glazed.

RAKE
A tapering strip cut off the edge of a pane of glass which is then described as "with rake off one side", or "off two sides", as the case may be.

REBATES
The cut-away portion of the window frame forming an angle to receive the glass.

REVEAL
The flat sides of a window or recessed opening.

SADDLE-BAR
The iron bar, fixed between jambs or mullions, to which panels of leaded glass are tied to increase rigidity.

SANDBLASTING
A process whereby the surface of glass is obscured by means of sand, or other abrasive propelled in the form of a jet against the surface.

SATIN OR VELVET
FINISH
The surface produced on glass by more than one treatment with acid.

SHADED
SANDBLAST
Delicate gradations of surface obscuration obtained by sandblasting.

SHAPES
Any pane of glass which is not rectangular or square.

SHUT-OFF
A block of refractory material which floats on the surface of the molten glass and is used in conjunction with the tweel to separate the drawing kiln from the tank.

GLOSSARY

SIGHT SIZES	The sight size, also referred to as the "daylight size" of an opening admitting light is the size of the daylight opening. The sight size of a glazed picture, mirror or similar frame is the size of the visible glass area.
SILVERING	A process whereby silver is deposited on glass and coated with a suitable protective medium.
SKY LIGHT	A glazed frame in a roof, providing light and ventilation.
SPRIG	The glazier's sprig is a small, headless nail, used in addition to puttying, for securing large panes of glass to wooden frames.
SQUARES	The generic term for pieces of glass cut to sizes for fitting into window openings.
STAINED GLASS	A term loosely used to cover the craft of leaded glazing, the actual glass painting, as well as enamelling and kindred arts. See painting.
STIPPLING	A process whereby the surface of glass is obscured by treatment with a mixture of acid with an inert substance. The depth of penetration of the acid varies over the surface of the glass, producing a stippled effect.
STOCK SIZES	The sheets of glass supplied to the glass merchant for cutting up.
SUBSTANCE	The thickness or weight per square foot of glass employed for glazing purposes.
TRIPLE OR FRENCH EMBOSSING	An embossed surface of glass, having four or more shades of obscuration.
TWEEL	A block of refractory material suspended from above and allowed to rest on top of the shut-off.
VENT	A ventilator; steel-frame window manufacturers refer to all windows or parts of windows which open as "vents" and indicate them on drawings by diagonals.
VELVET FINISH	See Satin Finish.
WAIST	The tendency for any material in a semi-plastic state to narrow in the width when drawn in the form of a continuous ribbon.
WASHLEATHER GLAZING	Glass bedded in washleather and secured with beads.
WHITE ACIDING	The surface produced on glass by a single treatment with acid.
WINDOW OPENING	The aperture in the wall of a building into which the window is fitted.

INDEX

The list of contents on page 5 indicates the place of the various types of glass in the book: this index deals chiefly with references to quotations.

For Product Safety Concerns and Information please contact our EU
representative GPSR@taylorandfrancis.com
Taylor & Francis Verlag GmbH, Kaufingerstraße 24, 80331 München, Germany

* 9 7 8 1 0 3 2 3 6 5 8 1 7 *